Also by Helen Fisher

Why We Love

The First Sex

Anatomy of Love

The Sex Contract

Why Him?
Why Her?

Finding Real Love by

Understanding Your Personality Type

Why Him? Why Her?

HELEN FISHER

Henry Holt and Company
New York

Henry Holt and Company, LLC
Publishers since 1866
175 Fifth Avenue
New York, New York 10010
www.henryholt.com

Henry Holt® and *® are registered trademarks of*
Henry Holt and Company, LLC.

Distributed in Canada by H. B. Fenn and Company Ltd.

Library of Congress Cataloging-in-Publication Data

Fisher, Helen E.
  Why him? why her?: finding real love by understanding your personality
type/Helen Fisher.—1st ed.
        p. cm.
  Includes bibliographical references (p.).
  ISBN-13: 978-0-8050-8292-0
  ISBN-10: 0-8050-8292-1
  1. Personality—Physiological aspects.   2. Interpersonal relations—
Psychological aspects.   3. Mate selection—Psychological aspects.   I. Title.
  BF698.9.B5F57   2009
  155.2'64—dc22   2008033877

Henry Holt books are available for special promotions and premiums.
For details contact: Director, Special Markets.

First Edition 2009

Designed by Meryl Sussman Levavi

Printed in the United States of America

1   3   5   7   9   10   8   6   4   2

For Ray, Lorna, Audrey and the rest of my family

Somewhere there waiteth in this world of ours
For one lone soul, another lonely soul—
Each chasing each through all the weary hours,
And meeting strangely at one sudden goal;
Then blend they—like green leaves with golden flowers,
Into one beautiful and perfect whole—
And life's long night is ended, and the way
Lies open onward to eternal day.

—SIR EDWIN ARNOLD

CONTENTS

Why Him? Why Her?

I am large, I contain multitudes.
— WALT WHITMAN

EAVESDROPPING ON MOTHER NATURE:

Why Him? Why Her?

"Now you will feel no rain, for each of you will be shelter to the other. Now you will feel no cold, for each of you will be warmth to the other. Now there is no more loneliness for you. But there is one life before you. Go now to your dwelling place, to eat to your days together. And may your days be very long upon this earth."

The Apache Indians of the American Southwest probably recited this wedding poem for centuries before I heard it in La Jolla, California, in 2006. It was an early June evening, the sky still pink and blue, the sea smells wafting through the windows as I sat in a folding chair on the second story of a fancy Italian restaurant. An older gentleman was conducting a short wedding ceremony, one mixed with rituals from the Christian, Jewish and Apache traditions. And before me glowed the two celebrants, Patrick and Suzanne—one of the first couples to marry after meeting on the Internet dating site I had helped to design, Chemistry.com.

Patrick had been a journalist in New Orleans until he lost his job, his home and all of his belongings to Hurricane Katrina in 2005. West he went, taking up residence with relatives in Los Angeles in February 2006. Days after settling in, he joined Chemistry.com and received his first recommended match: Suzanne, a lawyer living in La Jolla. That first night they talked for three hours on the phone. They met the following weekend and fell passionately in love.

So on a balmy evening during an April vacation together in Paris, Patrick took her to the top of the Eiffel Tower and proposed. The dazzled young woman grinned her "yes." So here I sat at a fancy Italian restaurant in La Jolla, surrounded by some fifty of their friends and relatives on this festive wedding eve.

I like being around people who are in love. They have a contagious energy. This force was palpable in the groom, the first to arrive for the nuptials. He burst into the room, filling it with his vivacious charm. Although we had never met, he greeted me warmly. We instantly struck up a conversation about the evolution of the English language, his experience as a journalist in some dangerous parts of Asia and some of my past work on the brain chemistry of romantic love.

Others soon arrived, and we took our places on the folding chairs facing a small bar strewn with lilies. Last came the bride. I was stunned when saw her—a tiny, perfectly formed, porcelain-like doll, with huge blue eyes and long auburn hair in soft ringlets wreathed in forget-me-nots. Like the mythological Helen, Suzanne had a face that could launch a thousand ships. And her vigor matched his. She was enraptured by her prince, gazing at him and grinning with uncontainable effervescence as she said "I do."

Someone played a flute. The Apache poem was read. And as the bride and groom walked down the makeshift aisle between our seats, we blew bubbles at them from the little bottles left on our chairs. Then came the feast: platters of Cavatelli Marinara, Antipasto Rustico, mussels, sausages, Chicken Fra Diavolo—a host of Italian favorites appeared at every table amid the balloons, confetti and champagne as the disc jockey blasted out old tunes and we wildly danced. Patrick and Suzanne swirled among us, radiating joy.

"Love hopes all things," the Bible says. I hoped for Patrick and Suzanne. But I also had a reason to be optimistic about their marriage. I knew some things about their personalities because both had taken my personality test, a series of questions I had devised to establish some basic things about a person's biological temperament. Both had told me their test results. And from these data, I was confident that Patrick's particular chemical profile would complement Suzanne's, creating a biological and psychological cocktail that would keep them captivated with each other for years.

Temperament and Love

We have many inborn tendencies. Indeed, scientists now believe some 50 percent of the variations in human personality are associated with genetic factors. We inherit much of the fabric of our mind.

But what is personality?

Psychologists define it as that distinct cluster of thoughts and feelings that color all of a person's actions.

Your personality is more than just your biology, of course. Personality is composed of two fundamentally different types of traits: those of *character* and those of *temperament*.

Your character traits stem from your experiences. Your childhood games; your parents' interests and values; how people in your community express love and hate; what relatives and friends regard as polite, dangerous or exciting; how they worship; what they sing; when they laugh; what they do to make a living and relax—these and innumerable other cultural forces combine to build your unique set of character traits.

The balance of your personality is your temperament, all of the biologically based tendencies you have inherited, traits that emerge in early childhood to produce your consistent patterns of feeling, thinking and behaving. As the Spanish philosopher José Ortega y Gasset put it, "I am, plus my circumstances." Temperament is the "I am," the foundation of who you are. Curiosity; creativity; novelty seeking; compassion; cautiousness; competitiveness: to some degree, you inherit these and many other aspects of your disposition.

It is this part of the human spirit I had examined in Patrick and Suzanne—their biological temperament.

Born "Me"

No one knows precisely how many traits of temperament we human beings inherit. But studies of identical twins suggest we inherit many. Take the "giggle twins," as they were called by staff members of the Minnesota Twin Study in the 1970s because these women would erupt with peals of laughter at the slightest jest or odd turn of phrase.

Daphne and Barbara were born to an unmarried Finnish student living in England in 1939. Barbara was adopted by an English groundskeeper who worked in a public park, while Daphne grew up in the home of a wealthy metallurgist. Yet when they first came together again at age thirty-nine as part of the Minnesota Twin Study, which focused on identical twins reared apart, both loved good pranks and both had giggled all their lives. Both regularly sat on their hands to keep from nervously gesticulating. Both had dyed their hair auburn. Both were effusively energetic. Both hated math and sports. Both avoided commercial television. Both preferred the color blue. Both were unwilling to give any political opinions. And both had met their husbands at age sixteen at a town hall dance and married in the autumn. Their IQ scores were nearly identical, too, despite Daphne's expensive education and Barbara's far more modest schooling.

Coincidence?

Psychologist Thomas Bouchard, director of the Minnesota Twin Study, unearthed so many stories like this one that in the 1980s he proposed that dozens of personality traits have a degree of heritability. Among those with the strongest genetic links, he reported, were traditionalism, the willingness to capitulate to authority, aggressiveness, the drive to lead and the appetite for attention. As he wrote in 1984, "Both the twin studies and the adoption studies converge on the surprising finding that common family environmental influences play only a minor role in the determination of personality."

In recent decades human behavior geneticists have added substantially to this list of traits linked with our DNA. More important to this book, scientists now know that groups of interacting genes influence behavior, even act together to create *behavior syndromes*. For example, if you have a biological appetite to seek novelty, you are also likely to be energetic, spontaneous, risk taking, curious and creative. If you are predisposed to be traditional instead, you are also likely to be loyal, cautious, respectful of authority and eager to make plans and follow schedules. We express constellations of related biological traits,[1] creating what are commonly called personality types.[2]

In fact, after doing extensive research on the biological underpinnings of personality types, I have come to believe that each of us expresses a unique mix of *four* broad basic personality types. Moreover, our primary personality type steers us toward specific romantic partners. Our biological nature whispers constantly within us to influence who we love.

These thoughts and more were swimming through my mind as I blew those bubbles at Patrick and Suzanne on that enchanting wedding evening. I thought both had found their soul mate.

Who are you? Why are we naturally attracted to particular mates? My investigation of these mysteries started over the Christmas holiday in 2004.

Match.com

"Why do you fall in love with one person rather than another?" This is what the executive team at Match.com wanted to know when I met with them two days after Christmas 2004 in New York City. Match .com is the world's largest Internet dating site. And I had been invited to spend the day with them, thinking. Midmorning, they asked me this fundamental question.

"No one really knows," I responded.

Psychologists have determined that men and women tend to fall in love with individuals from the same ethnic and socioeconomic background; with those of a similar level of intelligence, education and physical attractiveness; with individuals holding similar religious, political and social values; and with those who have a similar sense of humor. We also fall in love when the timing is right; and often with someone who lives or works nearby. Your childhood plays a huge role in your romantic choices, although no reliable patterns have ever been established. We tend to fall in love with someone who provides us with the things we need. And people often fall in love with those who are in love with them.

But, as I told the Match.com executives, how two individual *personalities* match up remains unknown. People do not necessarily court, live with or marry someone with similar *or* different personality traits. In fact, some 470 studies have examined the mesh of two personalities in a marriage. And psychologist Marcel Zentner summed up these data, saying, "Preference for similarity in personality characteristics varies substantially across traits and individuals." As he put it, "How two personalities may be best combined in a relationship remains at present an unresolved issue."

Yet your choice of mate will color every aspect of your life: your morning conversations in bed and at the breakfast table; your friendships, family reunions and weekend frolics; where you live; how you raise your children; most likely even your career. And certainly this choice will affect your tomorrows. Those babies you are likely to produce and send forth to multiply are your genetic future. Only a few times in your life will you mix your seed with that of another and pitch your DNA toward infinity.

So whom you choose matters.

In fact, I found it hard to believe that evolution would leave this decision entirely to our human whims. I suspected that psychologists had simply not looked for the underlying biological mechanisms that direct our romantic choices.

So when the folks at Match.com asked me to consider helping them develop a sister site for their Internet dating service, one designed for men and women interested in a long-term partnership, I said I would think about it during the festive midwinter lull.

The holiday season twinkled on. But on New Year's Day I realized I had to come to grips with this opportunity—a chance to apply the newest data in neuroscience to the essential question of who you love, perhaps even help people find "the one." So I sat down at my empty desk and pulled out a blank sheet of paper.

What did I know about personality?

The Biology of Personality

Dopamine. I began with this brain chemical because I had studied the activities of this powerful and ubiquitous neurotransmitter for several years.

On impulse, I listed some of the personality traits I knew were associated with specific genes in the dopamine system: the propensity to seek novelty; the willingness to take risks; spontaneity; heightened energy; curiosity; creativity; optimism; enthusiasm; mental flexibility. I decided to call those men and women who expressed the traits associated with this biology *Explorers*. Patrick, I would come to realize, had a good deal of the Explorer in him.

I drew another blank sheet of paper from my desk drawer. What else did I know about personality?

Well, individuals who have inherited particular genes in the serotonin system tend to be calm, social, cautious but not fearful, persistent, loyal, fond of rules and facts and orderly. They are conventional, the guardians of tradition. And because these men and women are also skilled at building social networks and managing people in family, business and social situations, I dubbed those who had inherited this constellation of genetic traits *Builders*.

I had also studied testosterone. Although testosterone is often associated with males, I knew that both men and women are capable of expressing particularly strong activity in this neural system. Moreover, those who inherit this chemistry tend to be direct, decisive, focused, analytical, logical, tough-minded, exacting, emotionally contained and good at strategic thinking. They get to the point. Many are bold and competitive. They excel at figuring out machines, mathematical formulas or other rule-based systems. Many are good at understanding the structure of music, too. I named these people *Directors*.

Last in my store of biological knowledge were some of the traits linked with estrogen. Women *and men* with a great deal of estrogen activity tend to see the big picture: they connect disparate facts to think contextually and holistically, expressing what I call *web thinking*. They are imaginative. They display superior verbal skills and excel at reading postures, gestures, facial expressions and tones of voice, known as *executive social skills*. They are also intuitive, sympathetic, nurturing, mentally flexible, agreeable, idealistic, altruistic and emotionally expressive. I christened the people of this broad biological type *Negotiators*.

Other chemical systems play a role in personality, of course. We may have as many as a hundred different kinds of neurotransmitters (smaller molecules) and some fifty types of peptides in the brain. But most keep the heart beating or orchestrate other basic functions. It is increasingly apparent that these four chemicals—dopamine, serotonin, testosterone and estrogen—play lead roles in producing aspects of personality.

Two others should be mentioned, though. Norepinephrine, a chemical closely related to dopamine, undoubtedly contributes to some of the Explorer's traits, especially their energy and impulsivity. And oxytocin—a chemical synthesized, stored and triggered (in large part) by estrogen—most likely plays a role in the Negotiator's compassion, nurturing, trust and intuition. In fact, *families* of chemicals produce the Explorer, Builder, Director and Negotiator. The specific activities of any one chemical are not as significant as the *ratios and interactions* among all of them and several other neural systems.[3,4]

Nevertheless, only dopamine, serotonin, testosterone and estrogen have been directly associated with a wide range of personality traits. So variations in these four chemicals most likely form the foundation of these four basic styles of thinking and behaving.

But does your personality type guide who you love?

I decided to find out.

The Personality Type Study

I accepted the consulting job as scientific adviser to what would become the new Internet dating site Chemistry.com. And in January 2005, I began to create the core statements in the questionnaire members would fill out when they joined this service, a questionnaire to establish their personality type. I based all my queries on genetics and neurochemistry.

For example, I posed the statement "I do things on the spur of the moment." I reasoned that Explorers would be most likely to "strongly agree" because certain genes in the dopamine system are associated with impulsivity.

"My friends and family would say I have traditional values." Because the drive to follow social norms is a hallmark of the serotonin system, I thought Builders would be particularly partial to this statement.

To establish the degree to which a member is a Director, I offered the statement "I am more analytical and logical than most people." High-testosterone men and women, I believed, would feel this described them accurately.

And because Negotiators tend to be highly imaginative due in large part to estrogen activity, I included the statement "I vividly imagine both wonderful and horrible things happening to me."

Then, with psychologists Jonathan Rich and Heide Island,

whom I invited to join me on this project, I perfected the questionnaire. It was launched nationally as part of Chemistry.com in February 2006. The point of this test: to measure your biological profile for
dopamine, serotonin, testosterone and estrogen—or the degree to
which you are an Explorer, Builder, Director and Negotiator.[5,6]

As I write this paragraph, some five million Americans have
filled out the questionnaire, along with another 1.8 million in thirty-
four other countries, because this personality test forms the basis of
the more extensive questionnaire and matching system used by
Chemistry.com. And upon reviewing the data in a sample of 39,913
anonymous members of Chemistry.com, I vividly saw how differently
each of these four personality types thinks and acts.

But I wanted to know more: Does each personality type like to
do something different on Saturday night? What are their religious
and political preferences? Do they pursue different vocations? Where
does each type prefer to live? What kinds of friends does each type see
regularly—a social crowd, intellectuals, adventurers or activists? How
often does each type fall in love? What are their views on kissing in
public? Do they regard sex as an important part of marriage? How
much personal autonomy and closeness does each type need in a
partnership? I was even curious about the state of their sock drawer:
messy or neat?

So I also asked these and other questions and examined the responses in this same sample of 39,913 anonymous members of
Chemistry.com. The group's average age was thirty-seven; 56.4 percent were women; and 89.6 percent were heterosexual. Explorers
were 26 percent of the sample, while Builders were 28.6 percent, Directors were 16.3 percent, and Negotiators were 29.1 percent of the
population.

Sure enough, each broad personality type does fraternize with a
different crowd, wants to live in a different place and has different religious values. These four types even doodle differently. And birth order
makes no difference in one's personality type, one of many indications
that I was measuring four different temperament types—styles of

thinking and behaving that stem from our biology and are soft-wired in the brain long before we leave the womb.

All the statistical data collected on this sample of 39,913 anonymous men and women, what I call the *Personality Type Study*—as well as information from genetics, neuroscience, anthropology, psychology and other scientific disciplines—form the basis of my understanding of Explorers, Builders, Directors and Negotiators.

The Mate Choice Study

Next, I needed to find out if a person's biological temperament steers him or her toward a particular personality type as a romantic partner. I had some evidence that each type would be particularly suited to a specific other. Nevertheless, I did a second study, the *Mate Choice Study,* using statistical data on a different sample: 28,128 anonymous heterosexual members of Chemistry.com.

When I examined whom these men and women *chose* to meet, I saw nature's plan: Explorers are attracted to other Explorers—people with many *similar* traits of temperament. Builders also gravitate to people like themselves, other Builders. Directors, however, gravitate to Negotiators. And Negotiators are drawn to Directors. These two personality types are attracted to individuals with a *complementary* temperament.[7] (See Appendix.) Moreover, these patterns occur whether one is a male or female.

No wonder so many scientists and laymen think that "opposites attract" while so many others believe "birds of a feather flock together." Both patterns occur—depending on your primary personality type. I felt as if I had sneaked into Mother Nature's kitchen and stumbled on her recipes for who we love.

But how does an individual on an Internet dating site scan through an array of photographs and essays (profiles) and select, even recognize, individuals of a particular biological type?

As it turns out, our faces and our words say much about who we are.

What's in a Face?

Take the iris of the eye. The crypts (or pits) of the iris and the furrows (the lines curving around the outer edge of the iris) are linked with specific traits of temperament. People with more furrows are more impulsive, as well as more willing to indulge their cravings—traits of the Explorer. Individuals with more crypts, however, are more trusting, warmhearted and tender—traits of the Negotiator.

Perhaps this is, in part, why we stare into the eyes of a potential lover, even in a photograph. We are unconsciously picking up subtle messages about temperament—how he or she is likely to think and behave years down the road.

We signal even more about our temperament with the structure of our face. Chiseled and square jaws, high and prominent cheekbones, heavy brow ridges and broad high foreheads are all signs of testosterone activity, the primary chemical of the Director. Even women with elevated testosterone display many of these traits. Just look at the high cheekbones, broad chins and high foreheads of many female celebrities.

Both sexes also signal estrogen levels with their face. Clear smooth skin, full lips, small noses, round soft faces, delicate brow ridges and other babylike facial features all indicate high estrogen activity.

I have found no studies that report on the facial traits associated with dopamine, the dominant chemical of the Explorer, or serotonin, the primary chemical of the Builder. With one exception: individuals with specific dopamine activity tend to have more expressive faces. Nevertheless, I suspect Explorers and Builders both have distinctive facial signals—and someday scientists will discover them.

"Promise, large promise, is the soul of an advertisement," Samuel Johnson, the eighteenth-century literary lion, said. We promise much with our faces, including some of the fundamental traits of temperament we bring to a partnership.

The Lure of Words

Explorers, Builders, Directors and Negotiators also signal their particular temperaments with their words.

This proposition has a history, known as the *Lexical Hypothesis*.
In the 1930s, psychologists proposed that when individuals describe
themselves they choose words and phrases that emphasize traits they
regard as central to who they are. With time, these words become en-
coded in their speaking habits.

So using another statistical sample, this time of 178,532 anonymous men and women on Chemistry.com, I launched my *Word Type
Study*. I composed a list of 170 words and examined how often each
personality type used these words as they described themselves and
what they were looking for in a mate.

Adventure is the word most often used by Explorers. The other
nine of their top ten most-used words (in descending order) are *venture, spontaneity/spontaneous, energy, new, fun, traveling, outgoing, passion*
and *active*.

Family is the most-used word among Builders. Following this,
Builders are particularly partial to the words *honesty, caring, moral/
morals, respect, loyal, trust, values, loving* and *trustworthy*.

Intelligent and *intelligence* together top the list of words used by
Directors. But they also favor *intellectual, debate, geek, nerd/nerdy, ambi-
tion/ambitious, driven, politics, challenge/challenging* and *real*.

Can you guess the favorite words of Negotiators? *Passion* and
passionate. These men and women also prefer *real, heart, kind/kind-
ness, sensitive, read/reader, sweet, learning/learn, random* and *empathetic
/empathy*.

Words, words, words. As Mark Twain said, "The difference between the right word and the almost right word is the difference
between lightning and the lightning bug." Words are verbal electricity.
As members of Chemistry.com read the essays of potential partners,
they apparently see specific words that instantly draw them to particular personality types.[8]

Your Primary *and* Secondary Type

We are billboards of our genetic dispositions. And although each of us is a unique blend of all four personality types, we express some of these types more regularly (and more naturally) than others. I, for example, am predominantly an Explorer and secondarily a Negotiator. I have a few traits of the Director and even fewer traits of the Builder; and at times I can act "out of character."

But I have come to think that *both your primary and secondary biological type are central to your temperament.*

Patrick, Suzanne's new husband, certainly advertised his primary and secondary types: he is primarily a Negotiator and secondarily an Explorer. And Patrick showed many of the qualities of both these types minutes after he bounded into that Italian restaurant on his wedding night.

Foremost, Patrick was instantly likable; he had a fine-tuned sensitivity to those around him and a superb talent for handling people. He radiated warmth and authenticity. Moreover, Patrick "cared." He spent his spare time teaching English to immigrants from Africa and Asia and sending supplies to war refugees in foreign camps. Indeed, his verbal skills, his easy way with people, his compassion and his mental flexibility enabled him to thrive in countries where few dare to work and many fail who do. Patrick was a Negotiator through and through.

But his secondary type, the Explorer, was also highly visible. Patrick was infectiously spontaneous. When asked during the wedding ceremony to sip some wine, he good-naturedly responded, "All of it?" During one of the rock numbers, he grabbed the mic and did an Elvis Presley imitation in his stately white wedding suit—the antics of the true social risk taker. And all evening, he wove through the balloons to every table and dragged complacent "singles" out to swing to the rock 'n' roll. Patrick was a fine blend of the verbal, imaginative, compassionate Negotiator and the energetic, novelty-seeking Explorer.

Suzanne must have sized him up as soon as she saw his photo and read his essays on Chemistry.com.

I didn't spend much time with Suzanne on her wedding night; she was understandably very busy. But she had told me she was primarily a Director and secondarily a Builder. So I could make some educated estimations about her temperament.

As a Director, Suzanne was most likely more focused than her new husband; she probably shot straight for her goals in a determined yet inventive way. Along with that gorgeous face and hourglass figure, Suzanne was almost certainly more forthright, analytical, exacting and autonomous, too. And as a Builder, she was probably more conscientious and cautious than her new mate.

The "Right" Fit

Most remarkable about that lovely wedding, Patrick and Suzanne fit one of nature's patterns perfectly. As you recall, their primary types, the Negotiator and Director, gravitate to each other. Moreover, their secondary types, the Explorer and Builder, can balance one another, each providing strengths the other needs. So as I watched Patrick and Suzanne exchange their vows, then twirl around the dance floor, I found myself imagining how they would get along in the years ahead.

Would they still be in love fifty years from now?

I thought they would—for biological reasons.

Negotiators and Directors are very different in basic temperament. But these types can appreciate each other's mode of thinking. Both like to imagine possibilities and generate theories. So Patrick and Suzanne will probably enjoy talking together for years about their insights and personal philosophies. Yet each will bring complementary skills to their conversations. While Patrick will see the big picture, Suzanne will probably know more about a smaller piece of the social, political or ethical puzzle. Each will almost certainly impress the other with their knowledge and ideas.

These two types value their independence, although Suzanne is likely to be far better at expressing her autonomy since Directors tend to be forthright and tough-minded. But Patrick will appreciate Suzanne's frank, decisive nature—because Negotiators ruminate. They tend to live in an inner world full of soul-searching, options and ethical knots that keep them from making up their minds.

There will be mutual admiration, too. Suzanne is likely to admire Patrick's imagination and insight and his ability to understand and deal effectively with people. She will also bask in his compassion, exuberance and sincerity. And her skepticism will balance Patrick's easy trust.

But life won't be all roses for this dynamic couple. Directors admire self-control; so at times Suzanne may think Patrick's impulsivity is a bit over the top. She may become impatient with his intuitive side, too. Directors are logical, and the Negotiator's flights of imagination can leave them feeling ungrounded. Meanwhile, Patrick may occasionally find Suzanne's pragmatism maddening.

Their secondary types, the Explorer and Builder, will also have an impact on their relationship: Patrick may occasionally feel hemmed in by Suzanne's caution, while Suzanne may become annoyed when Patrick is impractical.

Nevertheless, nature has given Patrick and Suzanne many complementary primary and secondary traits. With this big boost from their biological dispositions, along with a bit of work and some luck, they have an excellent chance of remaining forever "in sync," just the way they were as they swung around the dance floor that exhilarating wedding night: two very different individuals in perfect step with each other.

Our "Other Half"

In *The Symposium,* Plato describes a dinner party in Athens in the home of Agathon in 416 BC. Many of the pundits of classical Greece were in attendance. And as they reclined on their couches, one guest suggested they entertain themselves by describing the God of Love. All agreed, and each took his turn eulogizing Love.

It was at this gathering that Aristophanes spun his well-known tale of ancient men and women. Each couple, he mused, was originally a single entity. Two heads; two torsos; two sets of genitals; four legs: each man-woman was one beast. But these giants became obstreperous. And because they threatened to assume the throne of Zeus, this master of the universe retaliated, splitting each giant into halves—creating man and woman. Ever since, Aristophanes declared, each of us has been searching for our missing other half.

What are your primary and secondary personality traits?

Please take my test in the following chapter so I can begin to offer you some of nature's insights on who you are, why you are attracted to one person rather than another and how to find, understand and keep your other half.

CHAPTER 2

FINDING YOUR PERSONALITY TYPE:

The Personality Type Test

This test comprises a series of statements. Read each statement and circle the answer that best applies to you:
SD (Strongly Disagree), D (Disagree), A (Agree), or SA (Strongly Agree).
(*Scoring instructions are at the end.*)

SCALE 1

1. I find unpredictable situations exhilarating.

     SD   D   A   SA

2. I do things on the spur of the moment.

     SD   D   A   SA

3. I get bored when I have to do the same familiar things.

     SD   D   A   SA

4. I have a very wide range of interests.

     SD   D   A   SA

5. I am more optimistic than most people.

     SD   D   A   SA

6. I am more creative than most people.

     SD   D   A   SA

7. I am always looking for new experiences.

     SD   D   A   SA

8. I am always doing new things.

     SD   D   A   SA

9. I am more enthusiastic than most people.

     SD   D   A   SA

10. I am willing to take risks to do what I want to do.

     SD   D   A   SA

11. I get restless if I have to stay home for any length of time.

     SD   D   A   SA

12. My friends would say I am very curious.

     SD   D   A   SA

13. I have more energy than most people.

     SD   D   A   SA

14. On my time off, I like to be free to do whatever looks fun.

     SD   D   A   SA

TOTAL _____

SCALE 2

1. I think consistent routines keep life orderly and relaxing.

     SD   D   A   SA

2. I consider (and reconsider) every option thoroughly before making a plan.

     SD   D   A   SA

3. People should behave according to established standards of proper conduct.

     SD   D   A   SA

4. I enjoy planning way ahead.

     SD   D   A   SA

5. In general, I think it is important to follow rules.

     SD   D   A   SA

6. Taking care of my possessions is a high priority for me.

     SD   D   A   SA

7. My friends and family would say I have traditional values.

     SD   D   A   SA

8. I tend to be meticulous in my duties.

     SD   D   A   SA

9. I tend to be cautious, but not fearful.

     SD   D   A   SA

10. People should behave in ways that are morally correct.

     SD   D   A   SA

11. It is important to respect authority.

     SD   D   A   SA

12. I would rather have loyal friends than interesting friends.

     SD   D   A   SA

13. Long established customs need to be respected and preserved.

     SD   D   A   SA

14. I like to work in a straightforward path toward completing the task.

     SD   D   A   SA

TOTAL _____

SCALE 3

1. I understand complex machines easily.

   SD   D   A   SA

2. I enjoy competitive conversations.

   SD   D   A   SA

3. I am intrigued by rules and patterns that govern systems.

   SD   D   A   SA

4. I am more analytical and logical than most people.

   SD   D   A   SA

5. I pursue intellectual topics thoroughly and regularly.

   SD   D   A   SA

6. I am able to solve problems without letting emotion get in the way.

   SD   D   A   SA

7. I like to figure out how things work.

   SD   D   A   SA

8. I am tough-minded.

   SD   D   A   SA

9. Debating is a good way to match my wits with others.

   SD   D   A   SA

10. I have no trouble making a choice, even when several alternatives seem equally good at first.

    SD   D   A   SA

11. When I buy a new machine (like a camera, computer or car), I want to know all of its technical features.

    SD   D   A   SA

12. I like to avoid the nuances and say exactly what I mean.

    SD   D   A   SA

13. I think it is important to be direct.

        SD   D   A   SA

14. When making a decision, I like to stick to the facts rather than be swayed by people's feelings.

        SD   D   A   SA

TOTAL _____

SCALE 4

1. I like to get to know my friends' deepest needs and feelings.

        SD   D   A   SA

2. I highly value deep emotional intimacy in my relationships.

        SD   D   A   SA

3. Regardless of what is logical, I generally listen to my heart when making important decisions.

        SD   D   A   SA

4. I frequently catch myself daydreaming.

        SD   D   A   SA

5. I can change my mind easily.

        SD   D   A   SA

6. After watching an emotional film, I often still feel moved by it several hours later.

        SD   D   A   SA

7. I vividly imagine both wonderful and horrible things happening to me.

        SD   D   A   SA

8. I am very sensitive to people's feelings and needs.

        SD   D   A   SA

9. I often find myself getting lost in my thoughts during the day.

        SD   D   A   SA

10. I feel emotions more deeply than most people.

      SD   D   A   SA

11. I have a vivid imagination.

      SD   D   A   SA

12. When I wake up from a vivid dream, it takes me a few seconds to return to reality.

      SD   D   A   SA

13. When reading, I enjoy it when the writer takes a sidetrack to say something beautiful or meaningful.

      SD   D   A   SA

14. I am very empathetic.

      SD   D   A   SA

TOTAL _____

Scoring

To score the test, give yourself 0 points for each *SD* answer, 1 point for each *D,* 2 points for each *A* and 3 points for each *SA*. Then add up the number of points in each of the four scales; but *do not* add all the points of all scales together.

Scale 1 measures the degree to which you are an Explorer. Scale 2 measures the degree to which you are a Builder. Scale 3 measures the degree to which you are a Director. Scale 4 measures the degree to which you are a Negotiator. The scales with the two top scores are your primary and secondary personality types. For example, if you have 32 points on Scale 1, 20 points on Scale 2, 27 points on Scale 3 and 19 points in Scale 4, you are an EXPLORER/Director.

O wad some Power the giftie gie us
To see oursels as ithers see us!

— ROBERT BURNS

THE HAND YOU'RE DEALT:

Nature's Dice

S o, who are you?

"Nobody knows what's in him until he tries to pull it out," wrote Ernest Hemingway. What have you pulled out about yourself after taking this personality test? Are you a BUILDER/Negotiator? A DIRECTOR/Explorer? Or some other combination? No mix is better or worse, of course. In fact, we are all a blend of all four types; we simply measure higher on some of these scales than others. But I have come to believe your top two personality types generally dominate your personality.

This is not to say that your biology is your fate; we are not puppets dangling on a string of DNA. In fact, before I tell you more about this personality test and who you are, I must emphasize that your temperament goes hand in hand with all of your life experiences to make you who you are.[1]

This I am sure of—because I have an identical twin sister, Lorna.

We inherited the same biology, and we are both EXPLORER/Negotiators. Yet our individual experiences have sculpted us into different human beings.

To begin with, we can both tolerate risk. Lorna is a hot-air balloon pilot. She rides the air currents above the mountains of Colorado and Switzerland with only an oversized Bunsen burner to control her waist-high basket. As an anthropologist, I discuss in public such prickly topics as sex, adultery, divorce and gender differences. So we probably both express the same dopamine activity, the chemical associated with taking risks that predominates in the Explorer.

Lorna is also an oil painter. And although this profession has no obvious similarities with mine as a writer, Lorna builds a canvas with thousands of tiny brushstrokes, while I write books with thousands of little words. We are both focused, and we both make our living by being creative—more signs of similar dopamine activity. We also both have jobs that require good "people skills," so we undoubtedly have the same levels of estrogen, the primary chemical of the Negotiator.

Moreover, neither of us works in the corporate world or armed services, environments where structure, rules and lines of command prevail. Instead, we both work for ourselves; we both work alone; we both keep flexible schedules, pursue self-made goals and have unconventional occupations. So most likely we both express low serotonin activity, the principal chemical of the Builder.

And we both have the same degree of sex drive, suggesting that we both inherited the same base levels of testosterone—the major hormone of the Director.

Nevertheless, my daily life as an anthropologist and writer is very different from Lorna's as a balloon pilot and painter. We wear different kinds of clothes, pursue different sports, read different kinds of books, even live in different countries. Lorna and I are not replicas of each other.

My point is that all Explorers, Builders, Directors and Negotiators are not alike. Some Explorers climb mountains; others are history buffs, antiques collectors, theatergoers or voracious readers. Some

Builders manage businesses; others become soldiers or accountants; others are priests, rabbis or imams. Some Directors work in a lab around the clock to find the cure for cancer; others design computer chips, work in an auto body shop or tinker in their garage. And some Negotiators lead countries to peace at the conference table; others teach kindergarten; others are active in social causes.

The test you've just taken measures your innate dispositions—the hand you're dealt. Upon this genetic foundation, you build your fuller personality with your choices and experiences.

Being Yourself

Nevertheless, people are inclined to think and act in broadly different ways—those of the Explorer, Builder, Director and Negotiator. And most of us predominantly express two of these personality styles.

There are exceptions, of course.

My friend Keith, for example, regularly displays traits from three of these four personality types. He is highly personable, as well as poetic with words and verbal images, all traits of the Negotiator. But he makes his living as a world-class neuroscientist, performing complex technical laboratory experiments in the manner of the Director. And his scientific theories, speeches and poems all demonstrate the curiosity, creativity, flair and daring of the Explorer.

Others express all four personality types almost equally. Take Jennifer, an editor at a major American publishing firm in New York. After finishing my personality test, Jennifer called to tell me her scores were almost identical on all four scales. "I must be terribly boring," she exclaimed. "Hardly," I replied. "You are simply well balanced." But even though her scores are very close, I suspect Jennifer's primary and secondary types are still the best indicators of who she is.

Most of us can also pretend to be people we are not. As Miguel exclaimed in Cervantes' *Don Quixote,* "I know who I am, and I know too that I am capable of being not only the characters I have named, but all the Twelve Peers of France, and all the Nine Worthies as well."

We are all capable of acting "out of character." But doing so is tiring. Explorers who crave novelty are likely to have difficulty maintaining a boring work regime or rigid social schedule indefinitely. Cautious Builders will not be comfortable leading a life "on the edge" for more than a few weeks or months. Decisive, tough-minded Directors are likely to become more and more impatient if forced to wait for others to make up their minds. And compassionate, nurturing Negotiators are not likely to enjoy a job where they have to be ruthless, such as torturing people to get information or choosing who will live and who will die during an evacuation.

Your temperament is not entirely malleable.

This became clear to me one evening shortly after I had made a speech in Washington, D.C. A woman from the audience came up to me and explained that she had been dating a man for three years but never really felt like herself when she was with him. She pretended to like military history and television sports to please him. Around him she also behaved with a zany, naive zest that she didn't recognize in herself. She managed to pull it off, she said, but her performances exhausted her. "When he leaves town on a business trip," she confided, "I slip back into my down-to-earth self again."

"Integrity simply means a willingness not to violate one's identity," wrote psychologist Erich Fromm. I suspect this woman will eventually become so uncomfortable that she will begin to behave like herself with her lover. And as she regains her integrity, her relationship will either mature or fail.

But no matter what happens to her relationship, she will probably like herself better.

Liking Yourself

I say this because I continue to notice that just about everyone who takes this personality test seems to be happy with his or her basic temperament. Explorers like being curious and spontaneous; Builders like being cautious, loyal and traditional; Directors like their decisiveness, logic and exactitude; Negotiators like their empathy, altruism and imagination. Eleanor Roosevelt said, "I think somehow we learn who we really are and then live with that decision." We more than live with it; we like it.

I first realized that people tend to like their particular temperament during an awkward moment in my office. A young woman had contacted me, looking for a summer job. She had just gotten her PhD in psychology from a local university, and she wanted to work with a dating service. In she marched one spring afternoon, a tiny, trim, alert young woman, all buttoned up in a business suit. She settled on my couch, and we began the interview. Within a few minutes I asked her if she had taken my personality test. She had.

"What is your primary type?" I asked.

"A Director," she replied.

"Oh," I responded. I suddenly had a queasy feeling. I had just described the Director's tendency to be direct, sometimes blunt, even rude. So I started to reassure her, saying that she was probably very socially savvy and nice to people.

There was a pause. Then she snapped, "I'm not *nice* to people. My sister is always so *nice* to everyone—even people she doesn't like. I think that's weak. I like to say what I mean."

I was aghast at my mistake. In my effort to ascribe to her some traits that I particularly admire, those of Negotiators like myself, I had offended this fierce young Director.

The Stability of Self

Will this feisty young woman still "say what she means" thirty years from now?

I suspect she will.

For decades, psychologists have been examining how much a person's personality changes over time. They measure such changes by a method called *test-retest*. They give the same test to the same person on several different occasions spaced over weeks, months or years;

then they compare these scores. The results? Most traits of temperament remain stable across your life.

Children and teenagers have more variable test-retest scores, probably because they tend to change their perception of themselves as more of their capabilities emerge. Scores of the elderly are also more changeable across time, perhaps because senior citizens have learned to adapt their personality to suit the occasion. Men and women who test most accurately and consistently on personality tests are those who are active and free to express their natural dispositions—most of them people in middle age.

Certain traits do tend to change over your life, however. Adolescents around the world tend to be more anxious, open to new experiences and extroverted, as well as less conscientious and agreeable. In their twenties, they begin to change. And by middle age, men and women in countries as different as England, Germany, the Czech Republic and Turkey become more conscientious and agreeable, and less extroverted, open to new experiences and anxious.

But even in one's middle years, people going through a life-altering experience can misrepresent themselves on a personality test. Marrying; divorcing; losing or winning an important job; retiring; the death of an important friend or family member; addiction; a serious illness: any of life's joys or traumas can alter who you think you are. Remarkably, life-altering events can even turn specific genes on or off, creating enduring changes in your personality.

Not always for the worse. Sometimes these events are so dramatic that they unleash undeveloped traits we never knew we had. This kind of transformation must have happened to a former landlord of mine, George.

George's Transformation

George was the worst kind of proprietor, miserly and disagreeable. He was so cheap that he refused to send up heat or hot water to my fourth-story walk-up. Just as irritating, when George painted the building's stairwell, he painted it only as far as his own landing on the second floor, leaving the two upper stories of the townhouse to continue peeling. Moreover, he constantly accused all the tenants of unfounded misdeeds.

He was at his worst one morning when he pounded on my door, then berated me for pitching chewed artichoke leaves into the narrow hallway and down the building's steep old stairs. To be sure, I had eaten an artichoke the night before, but I certainly hadn't strewn my garbage throughout the building. As George stood there fuming, I could see around him: yes, there were my artichoke leaves, along with lemon wedges and other familiar edibles, all scattered along the stairwell. But when I pointed out that his lazy son must have spilled the contents of my trash bag as he dragged it across the nails that protruded from the rotting steps, George sputtered with rage.

Soon after this ridiculous incident, George experienced a life-changing event: he got cancer. Remarkably, he emerged from this experience as an agreeable man. And although he never sent enough heat or hot water to my apartment or painted the stairwell past his own front door, he remained gentlemanly, even gracious, until I left the building many years later.

Which personality style was more natural to George?

I will never know. But unless you undergo a mind-bending event, you will probably continue to be your unique combination of Explorer, Builder, Director and Negotiator for much if not all of your life.

"Gaming" the Test

But can you "game" this personality test, altering your answers to become someone you would prefer to be?

This question has interested psychologists for decades. Most believe there are three ways to view your results on any questionnaire. The test may have captured who you really are, or who you think you are (your *self-concept*); or how you wish to be perceived by others (your *self-presentation*). All three types of results say interesting things about you. Nevertheless, evidence indicates that most questionnaires measure who you really are.

Interestingly, personality tests (like the one you just took) that use obvious, broad, simple statements, such as "I do things on the spur of the moment" and "I am tough-minded," are more trustworthy than those using more subtle queries. In fact, when researchers add questions that are designed to measure lying and posing, these additional queries do not improve the validity of the results.

National Character

How about national character? Will your nationality make you more of an Explorer, Builder, Director or Negotiator?

I intend to find out, because my personality test is currently being offered on the Match.com Internet site in thirty-four other countries. As I analyze these international data, I expect to see how culture influences temperament.

But another personality test has already shown that men and women living in Ethiopia, India, Indonesia, Lebanon, Estonia and forty-three other countries, speaking a total of twenty-seven different languages, *share* an array of personality traits—be they old or young, rich or poor, pink or brown. Your beliefs about family, God, sex and community and your taste in literature, music and food will vary depending on where, when and how you were brought up. But your degree of curiosity, creativity, traditionalism, calm, decisiveness and compassion are not likely to differ whether you live in Cairo, Tokyo or La Paz.

These four personality types—the Explorer, Builder, Director and Negotiator—stem from biology, not culture.

Animal Minds

Some of the core traits of these personality types are even seen in other animals—suggesting that aspects of these four basic behavioral styles come across the eons, up from creatures in our deep history.[2]

Take curiosity, a hallmark of the Explorer. When scientists put a Pink Panther doll near the open cages of several Great Tits, the European version of the chickadee, some hopped right out to investigate it while others lingered in their cages. You don't have to have a big human brain to be curious. Even some fish are more curious than others.

Conscientiousness, a core trait of the Builder, is also seen in other creatures. A stunning example is Safi, a German shepherd and longtime companion of the renowned primatologist Barb Smuts. Barb picked up Safi at an animal shelter while Safi was still a pup. Before taking her home, however, Barb stopped at a local park full of dogs. Safi immediately joined the rough-and-tumble play of the local pooch population, until a squirrel darted by. Immediately all dashed after it, heading for the woods. "Here, Princey!" "Come, Muffin!" various owners shouted. Barb hadn't named her new puppy yet, so she just screamed, "Come back!" Safi was the only dog that heeded, returning immediately to Barb's side. Safi proved to be conscientious all her life. If Barb took a nap while on a hike, for example, Safi refused to sleep beside her. She stood guard. Only after Barb woke would Safi rest.

Birds and mammals are not the only creatures with *individual* personalities. Some drosophila flies behave a bit like Directors, being far more aggressive than others of their kind. Scientists observed this phenomenon when they constructed a home for a community of these creatures, known as "fruit flies" because they dart around your fruit on the kitchen counter on summer days. This particular housing project consisted of a small walk-in closet with two little honeyed poles where the drosophila could alight and dine. It was a nice place to live, with room for every fly. Yet every morning when the researchers entered to examine who was doing what, they found a single male dominating one of the two feeding stations while all the rest of the flies crowded on the other.

And if you know one octopus, you don't know them all. Take "Leisure Suit Larry," a giant Pacific octopus that lived until recently in an aquarium in Seattle. Scientists named him for his sexual approaches to anyone who passed his tank. With him lived "Emily Dickinson," a painfully shy female octopus who perpetually hid behind the aquarium rocks, and the devilish "Lucretia McEvil," who regularly destroyed the water filter and knocked over all the rocks in the dead of night.

In the animal kingdom, these four personality types are probably best observed, however, among a group of chimpanzees living in a colony at the Arnhem Zoo in Holland. When the primatologist Frans de Waal arrived there in 1975 to study the chimps, there were a total of twenty-three males and females in the colony.

Dandy was the colony's Explorer, a true risk taker. Every time the group staged a breakout from their pen, Dandy led the way. De Waal called him "the mastermind behind the escapes."

Luit was particularly agreeable, a core trait of the Negotiator. He had "an open and friendly character and [set] great store by company," De Waal observed.

Yeroen was known for his ambition. "He keeps a close watch on his interests," wrote De Waal. "No one else is considered when he is pursuing his goal. He is a real go-getter." Like Yeroen, many Directors jeopardize their health, safety and family lives to get ahead.

And Mama was calm and managerial—a true guardian and peacekeeper. She enjoyed "enormous respect in the community," according to De Waal. Builders, the pillars of society, are found among chimpanzees as well as humans.

For centuries, people have known that animals of many species have distinct and unique personalities. They have bred birds, dogs and horses for particular personality traits. Indeed, if I could magically convert my personality test into animal languages and entice whales, penguins, mink, rhinoceroses and all our other furry and feathered relatives to take it, I am convinced I would find Explorers, Builders, Directors and Negotiators living in all of nature's wildest places.

Evolution of Personality

Aspects of these four broad temperament styles probably evolved long before the dinosaurs tramped the steamy earth.

Among the billions of creatures that lived across these eons, Nature sometimes favored individuals that explored their environment; at other times it selected for those who were more cautious or more aggressive or more socially adept. And gradually these core personality constellations—-the forerunners of our modern human Explorers, Builders, Directors and Negotiators—proliferated among species as different as the ancestors of fruit flies, octopuses and chimps.

Then, as our human forebears descended from the fast-disappearing trees of ancient Africa some 5 million years ago, those with different dispositions took on different roles in their tiny bands. The ancestors of contemporary Explorers roamed far into the dangerous grass, returning with meat, nuts or information. The antecedents of today's cautious Builders guarded the group and gradually built the rituals of tribal life. The predecessors of our modern mechanically minded Directors invented better spears and traps and calculated the coming of the rains and waning of the moon. And the forebears of today's imaginative and intuitive Negotiators held the group together with their social skills.

As our ancestors pooled their natural strengths, they lived another day, another generation. Variability within the group had important social and economic payoffs. So these four basic personality types prevailed.

Scientists have even proposed that a specific gene for novelty seeking—a trait of the Explorer—originated as a rare mutation that became more common because it was so useful to survival. To test this theory, they gave groups of men and women jobs to do as a team. Sure enough, teams with some members carrying this gene in the dopamine system for risk taking excelled. Risk takers, the scientists concluded, help a group survive.

Specific genes for many other traits must have been adaptive too. And over time, the biological underpinnings of these four broad personality types—the Explorer, Builder, Director and Negotiator—were passed across the eons to you and me.

Hippocrates, Aristotle, Jung and Myers

Indeed, psychologist David Keirsey reports that these four basic personality types have been noted in Western history since the time of the ancient Greeks.

Hippocrates, a Greek physician living between 460 and 370 BC, first proposed the concept of four broad temperament styles based on the "four humors."

Aristotle (384–322 BC) believed that humankind sought happiness in one of four ways: through sensual pleasure, what he called the *hedone;* by acquiring assets, the *propraietare;* in logical investigation, or *dialogike;* or in expressing moral virtue, or *ethikos.* Aristotle perfectly described core traits of the Explorer, Builder, Director and Negotiator.

Galen, a Roman doctor living in the second century AD, again defined the primary traits of these four types: *Sanguine* individuals, he believed, were optimistic, impulsive, energetic and fun-loving—traits of the Explorer. *Melancholic* individuals were thoughtful, ponderous and meticulous—much like Builders. *Phlegmatic* men and women were unemotional, rational and observant—common traits of the Director. And *choleric* individuals were passionate—a core trait of the Negotiator.

Since then, the sixteenth-century Viennese physician Paracelsus, the eighteenth-century German philosopher Immanuel Kant, and later philosophers and psychologists—including Rudolf Steiner, Alfred Adler, Erich Adickes, Erich Fromm and, most notably, Carl Jung—have also embraced the idea that each of us has inherited a specific constellation of biological traits.

The most extensive categorization of temperament comes, however, from the work of Isabel Briggs Myers and her mother, Katharine Cook Briggs. In the late 1940s these women began to develop what would become the world's most popular personality questionnaire, the Myers-Briggs Type Indicator, or MBTI. This test has now been translated into sixteen languages and taken by some 30 million people to cultivate self-knowledge and group communication.

Why is the Myers-Briggs test so popular?

Because it works, thanks to biology. Although Isabel Myers was unaware of it, most of her types fit the same four basic biological profiles that I've identified.[3] Then in the 1990s, a brilliant psychologist, David Keirsey, a protégé of Isabel Myers, simplified her schema to four basic personality types: the Artisan, Guardian, Rational and Idealist.

I was unaware of Keirsey's types when I outlined my own set of four personality styles while sitting at my desk that New Year's Day in 2005. Only later did I become aware of the striking similarities.

These four personality types are also represented in non-Western traditions. For example, several North American Indian tribes historically living on the Great Plains of today's Midwest believed in a sacred medicine wheel representing the circle of life. To the east soared the eagle, the symbol of vision and illumination. To the west was the bear, a steady, cautious creature that hunkered in caves and didn't roam. North was represented by the buffalo, the epitome of reason and wisdom. To the south was the mouse, symbolizing innocence and trust. Each of these creatures represents basic traits of either the Explorer, Builder, Director or Negotiator.

In short, animal behaviorists, physicians, philosophers and psychologists have been describing central aspects of these four personality types for over two thousand years. But I have had the advantage of twenty-first-century science, enabling me to link these four temperament constellations with their biological underpinnings.

Falling for "Him" or "Her"

So the personality test you've just taken examines these basic aspects of "who you are"—the part of you that began to evolve long before our first human ancestors strode the earth.

But why did our human forebears evolve the tendency to be far more *attracted* to some types than to others? As you know, Explorers are more attracted to other Explorers, and Builders gravitate to other Builders, while Directors are significantly more likely to choose to meet Negotiators, and Negotiators usually choose to meet Directors. Why Him? Why Her? This is what the folks at Match.com originally asked me over the Christmas holiday in 2004.

As I think about each of these patterns of attraction, I can see why they evolved. Take the attraction between the Director and the Negotiator. Directors are tough-minded, direct, decisive, focused, technically skilled and competitive—traits many Negotiators need. Negotiators, on the other hand, are socially skilled, talented with words, mentally flexible and compassionate, traits many Directors need. Furthermore, both types like to play with ideas and theories. Yet the Director can bring depth to these discussions, while the Negotiator offers the broad overview.

Interestingly, these types may even enrich each other's temperament biologically. When the assertive Director challenges the pacific Negotiator, the Director may trigger the release of testosterone in the Negotiator, thereby making the latter more assertive too. And when the more affectionate Negotiator strokes and hugs the Director, this physical contact may trigger the release of oxytocin (and thus also estrogen) in the competitive Director, perhaps making the Director more trusting, intuitive, compassionate and nurturing. Most important, when the Director and Negotiator pool their natural talents, they make a powerful team for rearing young.

But why would Builders seek other Builders? Probably because these men and women are so traditional, loyal and devoted to home and family. Can you think of a better combination for raising babies than *two* loyal and devoted parents? In fact, data show that these serotonin-rich men and women also marry people like themselves. I suspect that most of the world's fifty-year marriages are made by Builders who marry other Builders—it's a fine strategy for perpetuating your DNA.

But what about the attraction between two Explorers? Studies have shown that these dopamine-rich men and women marry people like themselves, not only in the United States but also in Germany and the Netherlands. Who is going to raise the baby if *both* partners constantly dash off to do something new, either physically or intellectually? Explorers also express more dissatisfaction with their relationships, and these sensation seekers are more likely to divorce.

The Explorer-Explorer match does not appear to be a good strategy for raising children. Yet here, too, nature has a plan. Since Explorers are more likely to divorce and remarry, they are also more likely to bear children with more than one partner. In fact, it's commonplace to encounter the man or woman who has married twice and had children with each spouse. I don't recommend divorce and remarriage, but there's genetic wisdom here. In deep history, famines and other catastrophes weeded out many family members, particularly children. If ancestral Explorers produced more variety in their young, some of these children would survive hard times—passing on their DNA.

Long ago, four very broad basic temperament types evolved among our forebears. Moreover, each type became particularly drawn to those who either complemented or accentuated their personality in ways that helped them rear their young. Directors and Negotiators pooled their very different resources. Builders capitalized on their strengths to produce many young with a single partner. And Explorers created variety in their offspring. All were successful ways to combine and spread one's genes. They still are.

There is undoubtedly far more to nature's scheme. Your secondary type also figures in your choice of a romantic partner. For example, I can see the impact of my secondary type in my own relationships. As you know, I am an EXPLORER/Negotiator. But I gravitate to EXPLORER/Directors. Look at the symmetry: As an Explorer, I have someone to go adventuring with, another Explorer. As a Negotiator, I have someone who urges me to be more tough-minded and decisive, a Director.

There's no bad match. Each type of partnership simply has different strengths and weaknesses—and a different combination of many essential traits.

Extroversion and Introversion

But before moving on to chapters on who you are and who you love, I should mention that I did *not* measure one trait you may regard as central to your personality and your relationship: your degree of extroversion. Why? Foremost, because I feel the concepts of extroversion and introversion are widely misunderstood.

Psychologists regularly regard extroverts as outgoing, gregarious, assertive, energetic, enthusiastic and talkative. These people, they say, feel more joy, more pleasure and more of the other positive emotions. Introverts are supposedly shy and reserved instead.

But this definition excludes an essential aspect of extroversion: how you get your energy. As Isabel Myers ingeniously put it, extroverts are focused on the external world; they get their energy from others. Introverts, on the other hand, are oriented inward. As they explore their inner universe of feelings and ideas, they become relaxed, rested and renewed. Despite her perceptiveness, Myers and her colleagues failed to measure this core aspect of extroversion and introversion in their famous test.

Nevertheless, I like Myers's original definitions of the extrovert and introvert; they agree with my own experience. My twin sister, Lorna, for example, is affable, as well as energetic, talkative and enthusiastic. She's a good mixer, she's comfortable with people, and she makes and keeps friends easily. But she doesn't gravitate to them in her spare time. On Saturday afternoon, she would rather ride the winds above Switzerland in her hot-air balloon, sail the Mediterranean in a tiny boat or hole up in her studio in southern France and paint. And in the evening she often goes to the theater or an art exhibit rather than a social gathering.

The great pianist Vladimir Horowitz apparently had a similar tendency. When asked his price to play the piano at a Paris gala and then preside at a grand reception attended by all the local notables, he reportedly replied, "My fee is $100,000: $5,000 to play the piano and $95,000 to attend the reception." Horowitz would probably have agreed with something Lorna once said to me: "I like people very much, but they wear me out."

If introverts can be socially savvy, extroverts can be socially awkward. Take Sally, a neuroscientist. Sally is self-conscious; she makes long pauses and blurts out non sequiturs as she speaks. Yet in her spare time she entertains at home or goes out with groups of friends. As Sally says, "I love people; I suck energy from them."

To my way of thinking, Lorna and Horowitz are sociable introverts, while Sally is a social yet bumbling extrovert.

Moreover, I suspect that each personality type expresses extroversion differently. Extroverted Explorers are probably flamboyant conversationalists—until the tedium of seeing the same people regularly drives them to fresh adventures. Builders probably have the most friends and acquaintances of the four types but avoid conversations that reveal their inner selves. Directors are not likely to be gregarious with a host of others because they seek deep, intense conversations; but when they do engage with someone they admire, they can be thorough, deep and enthusiastic. And Negotiators are so agreeable that they can probably behave like extroverts even when they would much prefer to have a quiet, confidential conversation with just one friend.

So I regard extroversion as too complex a trait to measure on a short test.

But what deterred me most from measuring this aspect of your temperament is the fact that extroverts aren't necessarily attracted to other extroverts; introverts aren't more attracted to other introverts; and extroverts and introverts aren't significantly more attracted to each other. This aspect of personality doesn't play a decisive role in romantic attraction. We choose our mates for other traits.

For the same reason, I didn't test your degree of neuroticism and anxiety. People don't seek partners with a similar or different degree of these traits either.

Nature's Dice

Novelist Henry James once wrote, "Never say you know the last word about any human heart." Each of us is a complex combination of all four of these basic styles of thinking and behaving. And some of us override our biology as we choose romantic partners. Nevertheless, Mother Nature has rolled her dice within each of us. In fundamental ways, we are who we are and we are naturally drawn to some people far more than others.

So let's "push off," in the words of poet Alfred, Lord Tennyson, and "follow knowledge like a sinking star." The next four chapters are devoted to who you are, your fundamental personality type. The rest of the book examines who you love and how to use nature's laws to find and keep "the one."

CHAPTER 4

Let us not follow where the path may lead.
Let us go instead where there is no path,
And leave a trail.

—JAPANESE PROVERB

DRINK LIFE TO THE LEES:

The Explorer

I ate a tarantula last winter. It was coated with tempura batter. Nevertheless, it was a tarantula, with all those hairy legs. I didn't eat the whole thing—just the head end; the rest I left on my plate. That same evening, I also ate a scorpion, a sliver of alligator, some bear meat and a few bites of kangaroo. Somehow I missed the rattlesnake, yak, South American rat, Indonesian cockroach and turtle. And I passed on the worms. Worms I could not do. Even in a canapé, a worm is a worm.

What was I doing? Well, I was dressed in my white evening dress, wearing my bear claw necklace with twenty-one claws that I had bought years earlier while living on the Navajo Indian reservation in Arizona. And like several hundred others dressed in "black tie," I was attending the annual gala of the Explorers Club at the Waldorf-Astoria Hotel in New York City. There we all stood, among the crystal chandeliers, thick red carpets and flowing champagne, eating "exotic appetizers."

For me, it was a once-in-a-lifetime event. Never again. I'm an Explorer. But tarantula hors d'oeuvres? The only thing that saved me was that I wasn't wearing my glasses.

The Sensation Seekers

"Two roads diverged in a wood, and I— / I took the one less traveled by / And that has made all the difference." Robert Frost's great poem has touched millions, perhaps because it invites us to reflect on our life decisions, triggering feelings of wistfulness. But to take the path "less traveled by" is also mysterious, adventurous and exciting—to an Explorer.

I understand this type. Someone once said to my twin sister, "You're the only person I know who can get excited when your car gets a flat tire." Not all Explorers get excited by a flat, of course. Some strike out for Mount Everest, the Amazon or another exotic realm; these people seek the thrills of physical adventure. Some teeter on the brink of bankruptcy as they invent the newest widget. Some gravitate to wild sex, gambling, alcohol or street drugs; they like being out of control. Some revel in bizarre music, exotic art, counterculture social activities or avant-garde ideas. And many simply collect stamps or antiques, stay at home and read, or take advantage of every theater, movie or musical event in town. Explorers seek adventures of the mind and senses.

Such was Helen Keller, a woman who overcame severe disabilities to explore the world. Born in 1880 and stricken by a fever at the age of nineteen months that left her blind and deaf, Helen turned into a savage little tyrant, staging tantrums for attention, sulking and rocking in her bed. She once locked her mother in the pantry for three hours as she sat gleefully on the porch steps, relishing the vibrations of her mother's pounding on the pantry door. As she later wrote of her early childhood, "I was strong, active and indifferent to the consequences. I knew my own mind well enough and always had my own way, even if I had to fight tooth and nail for it."

This would change when twenty-one-year-old Anne Sullivan came to work with Helen, then age seven, at her home in Tuscumbia,

Alabama. Sullivan was a disciplinarian. When Helen threw a tantrum at the breakfast table to scare her off, hurling food and herself around the room, Anne leaped on Helen and held her down. But the new teacher had more than fortitude. She had grown up in a poorhouse in Massachusetts and supported herself and her brother working with the blind and deaf. She was proficient in American Sign Language. And as soon as she arrived at the Keller household, she began to train the unruly child.

Every time Helen touched an object, Anne grabbed her other hand and spelled out on her palm in sign language the English word for the object—*d-o-l-l, h-a-t, c-u-p*—until Helen's "aha moment" finally arrived. Years later, Helen recounted this experience: "We walked down the path to the well-house, attracted by the fragrance of the honeysuckle with which it was covered. Someone was drawing water and my teacher placed my hand under the spout. As the cool stream gushed over one hand, she spelled into the other the word *'water,'* first slowly then rapidly. I stood still, my whole attention fixed upon the motion of her fingers. Suddenly I felt a misty consciousness as of something forgotten—a thrill of returning thought; and somehow the mystery of language was revealed to me. I knew then that 'w-a-t-e-r' meant the wonderful cool something that was flowing over my hand. That living word awakened my soul, gave it light, hope, joy, set it free."

Helen Keller was a true Explorer. This tall, passionate and poetic woman went on to graduate from Radcliffe College, write several books and many articles, and fight for women's suffrage, the blind, workers' rights and birth control. She "spoke" to audiences of thousands and met every American president from Grover Cleveland to Lyndon B. Johnson. As she wrote, "Life is either a daring adventure or nothing."

Dopamine:
The Chemistry of Sensation Seeking

Explorers express a constellation of related traits. They are intensely curious and unusually creative. They are restless, energetic and spontaneous, often impulsive. They are willing to risk a great deal to pursue their many interests, and they get bored easily when not absorbed in something that intrigues them. They tend to be optimistic, irreverent and autonomous. Explorers are adaptable; they can play many different roles. Most are liberal in their political views, flexible in their personal lives and generous with their money, time and ideas. And Explorers crave novelty.[1]

"All pleasure consists in variety," wrote Samuel Johnson. Explorers believe it. They search for new experiences, a trait that psychologists call *novelty seeking* or *sensation seeking*. And although they are not necessarily fond of risk, they are willing to take risks to enjoy new, intense and exhilarating adventures. In my Personality Type Study, only Explorers overwhelmingly agreed with the statement "I find unpredictable situations exhilarating."

This sensation-seeking disposition is largely inherited. In fact, some of the contributing genes involved have been discovered. A gene labeled DRD4 controls much of the dopamine activity in brain regions used for thinking, feeling and motivation. And a specific version of this gene is associated with several varieties of novelty seeking. Old or young; male or female; rich or poor; educated in the ivory tower or on the mean streets: people who have inherited this gene in the dopamine system have an appetite for variety.

Dopamine acts in tandem with several other chemicals, of course, including the closely related norepinephrine and lesser players such as testosterone and serotonin.[2] So if you are describing yourself on a dating site or in the personals and hope to impress a doctor or endocrinologist, you might compose your essay thus: "I am a ———-year-old female/male with elevated activity in the mesolimbic dopamine system; low MAOb on my blood platelets; high circulating testosterone; low serotonin in many limbic regions; and low norepinephrine in my cerebrospinal fluid."

But to keep it simple, just say, "I'm an Explorer."

In my Personality Type Study of 39,913 anonymous members of Chemistry.com, 26 percent were Explorers. Some 27.1 percent of men were Explorers, while 25.1 percent of women were Explorers.

This is a very small yet statistically significant difference: more men were Explorers. But had I been able to give a longer test to measure more of the ways women explore—by reading, attending cultural events and traveling to exotic places, for example—I suspect I would have found just as many Explorer women.

Nevertheless, the Explorers I studied have much in common. Among these traits, both male and female Explorers would prefer to live in America's big cities—where the action is.

Boredom Susceptibility

"Any idiot can face a crisis—it's this day-to-day living that wears you out," wrote the Russian playwright Anton Chekhov. Not only do Explorers seek novelty; they are poor at tolerating boredom. They experience what psychologists call *boredom susceptibility*. Repetitive experiences, routine jobs, tedious companions exasperate them. In fact, many Explorers like interruptions. These diversions energize them. Explorers avoid monotony at almost any cost.

One famous Explorer was the American president John F. Kennedy. As the former White House correspondent Nancy Dickerson said of him, "You could do anything to him—steal his wallet, insult him, argue with him—but to bore him was unpardonable." Probably many Explorers would agree with JFK on this point. In my Personality Type Study, only Explorers responded enthusiastically to the statement "I get bored when I have to do the same familiar things."

You can see this mental state in an Explorer's doodles. "Doodling is the brooding of the hand," wrote artist Saul Steinberg. Indeed, the Explorers I studied were significantly more likely than other personality types to choose abstract, free-flowing, asymmetrical, open designs that suggest action and movement—designs characteristically made by individuals who express a great deal of dopamine activity.

Intense and Sustained Energy

"I cannot rest from travel: I will drink / Life to the lees." So spoke Ulysses in Tennyson's great poem named for the fabled Greek warrior and wanderer. Intense energy is another core trait of the Explorer.

Ernest Hemingway was a man with extreme energy. He ran with the bulls in Pamplona, hunted lions on the Serengeti Plain of Africa, caught marlin off Key West, Florida, and was one of the most daring war correspondents of his time. How did Hemingway have the energy to do all of this and write some of the greatest fiction of the twentieth century? With dopamine. Explorers abound in it. In my Personality Type Study, Explorers were significantly the most likely to respond positively to the statement "I have more energy than most people." Explorers were also the most likely to go dancing, see a live band or host a party—and the least likely to see a movie at home.

Despite her enormous physical handicaps, Helen Keller rowed, canoed, swam, hiked, danced and rode horseback. In fact, passersby used to gawk at her with amazement as she galloped by them in New York City's Central Park. She played chess and checkers; she went to museums, the theater and art shows; and she read voraciously, as she put it, "in the happy-go-lucky way I love." One of Keller's biographers, Dorothy Herrmann, wrote of her, "Helen's immense vitality amazed and touched everyone who knew her."

Greatly elevated dopamine (and norepinephrine) activity, in conjunction with other brain aberrations, can also produce hypomania (mild mania) or even full-blown mania. Psychiatrist Kay Redfield Jamison vibrantly described her childhood hypomania in her book *An Unquiet Mind*. "I found myself, in that glorious illusion of high summer days, gliding, flying, now and again lurching through cloudbanks and ethers, past stars and across fields of ice crystals . . . I remember singing 'fly me to the moons' as I swept past those of Saturn."

Helen Keller wrote of her frenetic childhood fantasies, too. But unlike Jamison—who would descend into full-blown mania in her twenties—Helen was blessed with an exceedingly active yet well-balanced mind.

Impulsivity

"He who hesitates is lost." Explorers believe it. Generally speaking, Explorers are also impulsive and uninhibited. Builders keep themselves in check to do their duty; Directors are monitored by logic and reason; and Negotiators often restrain themselves for ethical reasons. But for Explorers, waiting is slow death. Their cravings overwhelm them, and they leap for what they want. Explorers in my Personality Type Study responded particularly enthusiastically to the statement "I do things on the spur of the moment."

This *disinhibition* is directly related to the activities of dopamine. Women, even little girls, exhibit a little less of this impulsivity, as a particularly fiendish test illustrated: Girls and boys as young as three were asked to hold a piece of candy on their tongue for as long as possible before chewing, sucking or swallowing it. The little girls showed more restraint.

No matter what your sex, however, if you have inherited the Explorer's disinhibition, you are more likely than other types to enjoy smoking, alcohol, drugs, gambling, risky sex or some other sort of spontaneous, exhilarating behavior. Helen Keller described this love of "letting go" when she went tobogganing: "What joy! What exhilarating madness! For one wild, glad moment we snapped the chain that binds us to earth, and joining hands with the winds we felt ourselves divine!"

Daring feats; spontaneous retorts; last-minute plans; moments of intellectual discovery: Explorers express their spontaneity in many ways. One impulsive moment probably changed the course of modern Russia. In 1991 hard-line Soviet leaders had launched a coup against the liberalizing president Mikhail Gorbachev while he was vacationing in the Crimea. In a flash of impulsivity, Boris Yeltsin, then a government official and later the president, leaped onto a tank and made an impassioned speech. His words rallied Russian troops to support Gorbachev and smash the coup—and helped to fuel the struggle to modernize Russia.

I wonder how many other Explorers have played pivotal roles in our deep human history.

The Creative Drive

"I am more creative than most people." The Personality Type Study showed that Explorers were far more likely than other types to "strongly agree" with this statement. Be it poetry, music, literature, sculpture, speech making, sharpshooting, theater, cooking, the mechanical and industrial arts or the art of the business deal, Explorers often create as naturally as eagles fly.

Known as the *creative drive,* this creativity was recently measured by a curious test. Participants were asked to solve a series of anagrams, scrambled letters that can be rearranged to form words. Among them was the anagram *MPXAELE.* Don't look down the page. What word do you see?

The answer is *Example.*

Now the question. Did you solve this anagram by systematically trying out different letter combinations, or did you have an "aha moment" when the answer suddenly became clear to you? If you came to the solution with sudden insight, psychologists report, you tend to be more creative. Equally interesting, those who solved this anagram by sudden insight and those who arrived at the answer by methodical analysis show different patterns of brain activity. I suspect they have different activities in the dopamine system, too.

Erich Adickes, an early twentieth-century German philosopher, called creative people Innovators. Helen Keller was highly creative with words and images. Describing some of her times at the seashore, she wrote, "I thought it great fun to sit on a big rock in my bathing-suit and feel wave after wave dash against the rock, sending up a shower of spray which quite covered me. I felt the pebbles rattling as the waves threw their ponderous weight against the shore; the whole beach seemed racked by their terrific onset, and the air throbbed with their pulsations. The breakers would swoop back to gather themselves for a mightier leap, and I clung to the rock, tense, fascinated, as I felt the dash and roar of the rushing sea!"

It is now known that verbal and nonlinguistic creativity and the generation of ideas are all linked with specific dopamine pathways, particularly activity of the DRD2 gene.

Ingesting drugs that boost dopamine activity can spark creativity, too. This happened to a woman with Parkinson's disease. Soon after she began to take L-dopa, she became a remarkably talented poet.

The ancient Greeks glorified this aspect of the Explorer in the deity Dionysus. Dionysus was the god of the vine, the immortal who could bring you freedom, merriment, courage and ecstatic joy—or brutality and ruin. Dionysus was also the god of inspiration. And to worship him, Athenians would assemble for a five-day holiday in the spring—not in the surrounding forests or sacred temples, but in their theaters. Here they enjoyed inspiring plays written in tribute to this muse.

Mythologist Edith Hamilton wrote of these festivals, "The greatest poetry in Greece, and among the greatest in the world, was written for Dionysus"—the god of creativity.

Too Much Creativity?

The Greeks also linked creativity with madness.

They weren't far wrong. Very creative people tend to have more chemical imbalances, addictions and mood disorders, and dopamine is regularly involved. This "taint of blood," as poet Alfred, Lord Tennyson described it, worried him all his life. Five of his brothers, as well as his father, grandfather, both great-grandfathers, son and grandson were manic-depressives, a disorder of the dopamine system. Tennyson himself escaped this mania yet reaped its fruit: abundant creativity.

German composer Robert Schumann was less fortunate. Schumann brimmed with creativity—when he was manic. But he rarely composed when gripped by one of his long depressions. Alas, he lost his battle with dopamine imbalance and died in an insane asylum.

Schumann was not alone. When Kay Jamison examined the lives of forty-seven well-known British painters, sculptors, playwrights and poets, she found that 30 percent had needed treatment for mood disorders, as opposed to 5 percent of the general population. Poets struggled the most: 50 percent of them needed considerable medical care.

"Human salvation lies in the hands of the creatively maladjusted," wrote Martin Luther King Jr. Poets, playwrights, philosophers, artists, adventurers of all kinds have unquestionably improved the lives of millions with their creative gifts.

Curiosity: "For Lust of Knowing"

Another hallmark of Explorers is their curiosity. Helen Keller was enormously curious. From the moment she joined humanity at the water pump, she needed to know. As she wrote, "I recall many incidents of the summer of 1887 that followed my soul's sudden awakening. I did nothing but explore with my hands and learn the name of every object that I touched."

With time, Keller would read English, French, German, Greek and Latin, travel to more than thirty-nine countries and study philosophy, history, the sciences and literature. She loved to learn. Ever since the age of seven, she wrote, "I have devoured everything in the shape of a printed page that has come within the reach of my hungry finger tips . . . Literature is my Utopia."

Explorers tend to have many interests. In my Personality Type Study, Explorers gave a particularly enthusiastic response to all three of these statements: "I have a very wide range of interests"; "I am always doing new things"; and "My friends would say I am very curious." Some Explorers collect coins, comic books or stamps. Some are walking encyclopedias of facts or tunes or figures. Some stroll regularly through the diverse neighborhoods of big cities, urban hiking. Some visit almost every country in the world. And many stay in school—to learn. In the Personality Type Study, the more education you had, the more likely you were to have a high score on the Explorer scale.

"For lust of knowing what should not be known / We take the Golden Road to Samarkand," wrote poet James Elroy Flecker. Regardless of the cost, the time, the danger or the hardship, Explorers must "know." When thinking about why Herodotus underwent such extreme hardships to write on the history and peoples of the ancient world, the Polish biographer Ryszard Kapuscinski wrote, "I think it was simply curiosity about the world. The desire to be there, to see it at any cost, to experience it."

Herodotus's curiosity gushed from a fountain of dopamine.

Focusing with Ease

Curiosity is biologically connected with another trait of the Explorer: focused attention.

Psychologists measure focus by the strength and persistence of your *orienting reflex*. This reflex comes in three varieties: *initial focused attention, divided attention* and *sustained attention*. Explorers are poor at sustaining their attention, unless they are especially interested. But when these sensation seekers are first exposed to a new stimulus, they exhibit a strong initial orienting reflex—they focus their attention rapidly and earnestly. I'm not surprised. Dopamine helps you remember and prioritize salient data, as well as focus intensely. As a result, an Explorer who is challenged at work or play can exhibit remarkable concentration.

Explorers also excel at dividing their attention, a talent that reminds me of Bill, a pal from my days in graduate school. In the evening, this medical man used to settle down with a good book while he listened to both the radio and TV. It was chaos and cacophony to me. But periodically he calmly switched his intense attention from one stimulus to another. Specific genes in the dopamine system (DRD2 and DRD3) can be thanked for this agility at focusing short term.[3]

This relationship between dopamine and focused attention became clear to me one day at Rutgers University. I was riding the campus bus when I got into a conversation with one of my students. Our class was over, grades had been distributed, and I had a hunch she trusted me. So I casually asked her what drugs students were taking these days to study.

"Ritalin," she replied.

I wasn't surprised. Although I am not in favor of using drugs to study, Ritalin activates the dopamine in the brain, enabling one to concentrate, at least briefly.[4]

Motivation

Dopamine also promotes motivation and goal-oriented behaviors. It provides incentive. In fact, reduced dopamine activity in particular brain regions produces *amotivation,* the lack of desire. But in the right proportions, elevated dopamine activity is linked with enthusiasm, energy, focus, assertiveness and the drive to achieve. As a result, most Explorers want to make their mark—in the theater, on the athletic field, with their novel, play or poem, or in the political arena or business world. Helen Keller, for example, was absolutely determined to go to college; nothing deterred her—not even the black silence that engulfed her.

Oddly, this brain chemistry of motivation may also explain a related trait of many Explorers: procrastination. Many Explorers are so busy with their many interests that they simply don't focus on critical deadlines until the last hour.

But there may be method to their madness. I, for example, rarely write a speech until the day before I deliver it. Sometimes I am still composing as I hear myself being introduced. I don't enjoy these close calls. But I have long suspected that as my heart pounds and I feel short of breath from panic, my feelings of danger and excitement are actually triggering a monsoon of dopamine in my brain, a flood that gives me energy, focus and ideas.

Adaptability: The Salamander

"It is not the strongest of the species that survive, nor the most intelligent, but the ones most responsive to change," Charles Darwin wrote. Explorers would applaud this view; they are mentally flexible and adaptable.

For example, Explorer men and women usually hold permissive political, religious and sexual views. They are more likely to be atheists or agnostics. And Explorers are the least judgmental of the four types. Paracelsus, the Renaissance physician, referred to this type of person as a salamander, capable of changing its skin color to blend with almost any environment. In fact, Explorers are often good mimics; they can adopt different personas to appear just like those around them.

They can also take criticism fairly easily. Explorers live in a world of live-and-let-live.

They are utilitarian, too. Explorers do what works. Theodore Roosevelt summed up the Explorer's attitude, saying, "Do what you can, with what you have, where you are." These men and women tend to remain cool in crises, and act freely, rapidly and tactically. Many even enjoy an emergency that requires them to work against the clock. Catastrophes give Explorers energy, focus and creativity.

Autonomy: "Don't Let Them Tame You!"

Dopamine probably also contributes to the Explorer's deep need for autonomy.[5] Explorers are unconventional and nonconformist; strict standards of conduct leave them cold. They are not bound by the rules and schedules of the Builder, the logic of the Director or the ethical knots of the Negotiator.

Instead, they trust their impulses. Explorers launch into projects that appear impossible; make deals that seem sure to fail; and accept tasks that appear crazy to those around them. Choreographer Isadora Duncan encouraged this attitude in her dancers, saying, "You were once wild here. Don't let them tame you!"

Explorers often can't be tamed. Many see life as a "crapshoot." When they fail, they are likely to just shrug it off, regarding their misfortune as an unlucky throw of the dice. They don't tend to moan about their losses (as Builders do), dissect their blunders (as Directors do) or worry about tomorrow's errors (as Negotiators do). And because Explorers can have superb timing, and cleverly take advantage of their opportunities, they often win.

"Good manners are made up of petty sacrifices," wrote Ralph Waldo Emerson. Explorers prefer not to make sacrifices; few see self-denial as a virtue. Instead, they want to be flexible, spontaneous and free. Even schedules and routines annoy them, as my Personality Type Study verified. When asked to respond to the statement "I think consistent routines keep life orderly and relaxing," Explorers scored far below other types.

Flexibility; energy; creativity; restlessness; novelty seeking; motivation: these traits tend to pay. Of the four basic personality types, Explorers—both men and women—have the highest incomes. "Nothing ventured, nothing gained." The old aphorism is true. Explorers often win big.

Extravagant Generosity

Explorers are generous, too. They value their friendships, at least when friends are interesting or suit their needs. And they can be extravagantly generous with them, often on the spur of the moment. Indeed, Explorers are impulsive buyers; they purchase without planning ahead—a trait associated with dopamine. And many will also "give the shirt off their back" to help a friend or even a stranger.

Helen Keller was certainly spontaneously generous. She was only ten years old when her beloved dog, Lioness, was shot and killed by a sheriff enforcing the leash law in a nearby town square. Helen said with her fingers, "I'm sure they would never have done it, if they had known what a dear good dog Lioness was." This comment was reported in the local newspaper and soon printed by many others. Within days, Helen's devotees began to raise money to buy her a new pup. But she refused the gift, requesting instead that the money be given to help a five-year-old deaf-blind boy in Pittsburgh.

Optimism: A Sunny Personality

Galen, the Roman physician, called this personality type the sanguines to accentuate their extreme optimism. Of the four types, Explorers are the most optimistic and enthusiastic, traits associated with dopamine. Low activity in dopamine circuits, on the other hand, is associated with *anhedonia*, the inability to experience physical or social pleasures.

Pleasure means different things to different people, of course. Sensation seekers often derive pleasure from highly arousing experiences, while those who score low on sensation-seeking scales tend to derive pleasure from doing quieter things, often familiar activities with familiar people. But ecstatic pleasure appears to derive, in large part, from dopamine.

This is certainly what drives cocaine addicts to take their drug: cocaine increases the activity of both dopamine and norepinephrine. But the euphoria, the "high," comes from stimulating the brain's dopamine receptors, not those for norepinephrine.

"Mirth, with thee I mean to live," wrote poet John Milton. Most people want to be happy. But many Explorers feel almost obliged to seek joy, particularly of the senses. Helen Keller, for instance, was driven to please her senses. One compelling incident in her childhood involved climbing a tree.

Helen had been caught in a tree during a violent thunderstorm some weeks before, and the mere thought of climbing another, she wrote, "filled me with terror." But one spring morning she smelled the perfume of mimosa blossoms, and she couldn't resist. "I felt my way to the end of the garden, knowing that the mimosa tree was near the fence. Yes, there it was, all quivering in the warm sunshine, its blossom-laden branches almost touching the warm grass . . . I made my way through the shower of petals to the great trunk and for one minute stood irresolute; then, putting my foot in the broad space between the forked branches, I pulled myself up . . . I had a delicious sense that I was doing something unusual and wonderful, so I kept on climbing higher and higher . . . I sat there for a long, long time, feeling like a fairy on a rosy cloud."

Like many Explorers, Helen could not contain herself; she seized this opportunity to enjoy her senses.

On the Dark Side: Negative Traits of the Explorer

Philosopher Immanuel Kant once opined, "Out of the crooked timber of humanity, no straight thing was ever made." So it is with all of us. Explorers are energetic, restless, impulsive, creative, curious, original, generous, optimistic, autonomous, liberal, untraditional, daring and interested in many things. But even appealing traits can turn sour.

Some Explorers are so irreverent, for example, that they appear cynical or glib. Because these men and women are independent-minded, they can—sometimes unwittingly—trample on the cherished rules and traditions of their relatives and friends. Their desire for novelty and adventure can lead to philandering. They run from boredom, and as a consequence they can become narcissistic or reclusive when forced into tedious social schedules. Because Explorers are so spontaneous, they can be unpredictable and unreliable. And their curiosity can override their sense of responsibility.

I have a friend like this. He is almost always outrageously late. I finally found out why. As he prepares to leave his house, a book, magazine or newspaper catches his eye. With his insatiable curiosity, he picks it up, becomes absorbed and loses track of time. Because he has little respect for schedules, and even less respect for those who abide by schedules, he is not motivated to be punctual, either.

Explorers also tend to be poor planners, perhaps because of their contempt for schedules. Explorers can be disorganized, too. They are likely to have piles of paper around their desk at work, and books, magazines, sports equipment or other artifacts of their interests scattered on their bedroom floor, on the backseat of their car, even in their purse and pockets. This has a silver lining, however. As British writer A. A. Milne put it, "One of the advantages of being disorderly is that one is constantly making exciting discoveries." As you might expect, disorderliness is associated with dopamine.

Some will also find fault with the Explorer's lack of introspection. Perhaps due to their intense curiosity about the world around them, Explorers seldom indulge in self-analysis. They look out, not in. So Explorers can appear emotionally shallow, particularly to Negotiators, who are highly introspective. Explorers can also lack the loyalty of the Builder, the directness of the Director and the sympathy of the Negotiator. Explorers can be aggressive, in part because dopamine triggers testosterone and suppresses serotonin, thus creating a cocktail of hostility. Explorers also tend to divorce more regularly than other types.

Psychoanalyst Erich Fromm was probably right when he proposed that this personality type has a tendency to be opportunistic. Explorers tend to score very low on four of six major components of agreeableness: altruism, compliance, modesty and straightforwardness. They score only slightly higher on trust and tender-mindedness. And Explorers can be devious as they maneuver to stay one step ahead of anyone who would divert them from their goals.

Explorers live in the here-and-now. They travel fast. And if you can't keep up, they go alone. Their motto: carpe diem—seize the day.

Possible Physical Problems

I suspect that Explorers are prone to specific mental diseases, particularly manic-depressive illness, now called bipolar disorder.

Psychiatrist Kay Jamison's struggle with bipolar illness began during high school, due to an imbalance in dopamine activity. And by age twenty-eight, as a new assistant professor of psychiatry at the University of California, Los Angeles, she was on her way to raw insanity. In her book she describes one evening outside the hospital where she worked: "I was on the run. Not just on the run but fast and furious on the run, darting back and forth across the hospital parking lot trying to use up a boundless, restless, manic energy. I was running fast, but slowly going mad." Eventually lithium, a drug that suppresses dopamine circuits, saved her life.

Explorers can also suffer from various addictions, including alcohol abuse, drug addiction and/or gambling—as a conservative businessman in his fifties discovered when he began to take a medication for Parkinson's disease, L-dopa. L-dopa elevates dopamine activity. And soon after he began his daily regimen, this formerly self-controlled man gambled away his family's savings. He also searched the Internet until he found his high school sweetheart, and began menacing her with sexual demands.

Impulsiveness, elevated sexuality and gambling have been reported so often among Parkinson's patients taking medications that trigger dopamine activity that clinicians are now being advised to warn them of these potential side effects.

Of course, not all Explorers are cynical, narcissistic, unpredictable, unsympathetic, disorganized, Machiavellian or perpetually aggressive. Few suffer from bipolar illness or addiction either. Instead, many are energetic, witty, spontaneous, optimistic, original and generous and have many interests and ideas. Moreover, Explorers' secondary traits always add to their personality. EXPLORER/Negotiators can be very nurturing and intuitive. EXPLORER/Builders can be exceedingly loyal and organized. And EXPLORER/Directors can be forthright.

The Explorer in Youth and Middle Age

"I circle around, I circle around, the boundaries of the earth, wearing my long-winged feathers as I fly." This Navajo shamanic song captures the core of the young Explorer—particularly men.

Young Explorer men score much higher than young women on the more physical forms of exploration, getting thrills from skydiving, skiing treacherous "black diamond" trails, spelunking or diving off cliffs into pools far below. This youthful male craving for physical adventure may be due to the relationship between dopamine and testosterone: these chemicals trigger each other. This chemical partnership also explains why men are particularly physically adventurous in their late teens when testosterone levels peak and elevate dopamine activity, as well as why their appetite for physical thrills declines with age, as testosterone levels decline, lowering dopamine activity.

But women in their teens and twenties are just as eager to chase new adventures of the mind and senses—especially stimulating music and visual arts, drugs and the intellectual and social fringes, such as palm reading, radical politics or avant-garde healing practices.

The lust for "action adventure" declines with age in both men and women as levels of dopamine decrease. But older sensation seekers continue their exploration by reading, traveling, going to the theater, movies, music, or other arts events, collecting memorabilia or pursuing other creative ways to satisfy their restless spirits. Moreover, middle-aged women begin to score higher than men in overall sensation seeking. They buy more books. And they support the travel and leisure industry—cajoling their partners to accompany them to the Great Wall of China, Victoria Falls, Patagonia or myriad local cultural events.

Middle-aged Explorers of both sexes also keep up with their friends—often other Explorers. In my Personality Type Study, Builders consorted with a "social crowd," Directors fraternized with "intellectuals" and Negotiators mixed with "activists." Explorers, however, befriended people like themselves: "adventurers."

Scientists report that we tend to choose our friends according to our genetically based temperament. All four types in the Personality Type Study certainly support this proposition.

Born Free

"Search, always go on searching." These were among Leo Tolstoy's last words to his followers on his deathbed. Explorers do search. They are born free. They want to "think outside the box," "rock the boat," "make waves," "court danger," "keep an open mind" and "seize the day."

No one is a pure Explorer, of course. We are all a boundless mix of genes, chemicals and experiences that make each of us unique. But Helen Keller captures the soul of the Explorer with her description of rowing: "Sometimes I go rowing without a rudder. It is fun to try to steer by the scent of water grasses and lilies, and of bushes that grow on the shore . . . I like to contend with wind and waves. What is more exhilarating than to make your staunch little boat . . . go skimming lightly over glistening, tilting waves."

Explorers "drink life."

CHAPTER 5

Be faithful to that which exists nowhere but in
yourself—and thus make yourself indispensable.

—ANDRÉ GIDE

A PILLAR OF SOCIETY:

The Builder

I was in a trendy hotel rooftop bar, the Beach, when I saw my first Builder in courting action. The evening was in full swing. Some two hundred singles were clustered at the long Hawaiian tiki bar, standing around the two cabanas and the fire pit, or sprawled on the gleaming white sand that stretched from one corner of the rooftop to the other—all beneath the cobalt firmament and surrounded by the towering glass buildings of downtown San Diego. It was a sandbox in the sky, a perfect setting for a singles event.

San Diego magazine was hosting this adventure, in conjunction with Chemistry.com, and I was there. The event would be a challenge. The MC was screaming into the microphone, announcing the beginning of the evening entertainment: a miniversion of *The Dating Game*. Two journalists from a local television station, a female weather broadcaster and a male sports anchor, had been invited to participate, and both were waiting in the wings.

The plan was simple: First the female weather broadcaster would perch on a high stool in the right cabana while three eligible bachelors sat on stools in the left one. A curtain would isolate the woman from the three men, but the crowd could see all participants. My mission: to watch the verbal exchange as the female weather broadcaster asked the three bachelors several questions and then to explain to the largely inebriated crowd why she chose the man she chose for a date that night. I had been told only one thing about this woman: she was primarily an Explorer and secondarily a Negotiator.

Out traipsed the weather anchor as the singles sipped their Grey Goose cocktails and cheered. She began. "How would you ride me if I were a surfboard?" she queried Bachelor No. 1.

I braced myself. The crowd roared. Bachelor No. 1 was slow to answer, mulling over the question. *He has lost already,* I mused; Explorers like quick repartee. Bachelor No. 2 responded next with something slick. But Explorers are not impressed with "slick." Partly because they can be slick themselves and can spot it fast in others. Moreover, Bachelor No. 2 lacked the warmth and authenticity that I felt would appeal to her Negotiator streak. Then Bachelor No. 3 responded.

"First, I will get you wet," he said.

The crowd went wild. *He's won,* I thought. Explorers tend to be sexy because dopamine activates testosterone, the hormone of sexual desire. And this guy had flair.

Bachelor No. 3 then delivered a playful rant about catching her on a wave and riding her tenderly until she peaked.

Now the audience was howling. And I became sure I knew her choice. Bachelor No. 3 had mixed spontaneous and unpredictable bawdiness with genuine care, playing directly to the Explorer's need for excitement and the Negotiator's need for authenticity.

I was right. She chose Bachelor No. 3.

That was a relief; my understanding of her biology had not let me down. But I had more to do. Next would be the male sports anchor, a Builder.

I suspected he would be relaxed, social, interested in home and family, community-oriented, conscientious and cautious, as Builders tend to be. Moreover, Builders take dating seriously; they are not amused by those who appear devious, grandiose or falsely sexual. So I had a hunch this man would choose the bachelorette who was the most down-to-earth.

In sauntered the male TV anchor. The three bachelorettes took their seats. First he asked each woman to say "hello." The crowd was riveted. Bachelorette No. 1 gave him a low, throaty "helllloooo." Bachelorette No. 2 topped her with "Helllloooo, Baby." The crowd swayed and sighed. Then Bachelorette No. 3 said, "Hi." No frills, no thrills: her salutation was short and formal.

I had a hunch she would be "the one." An Explorer would have been put off by her lack of playfulness. A Negotiator would have sought more warmth. A Director would have dismissed her as too conventional. But a Builder might be enticed; Builders are traditional.

Bachelorettes No. 1 and No. 2 continued to respond to his queries with quick, self-revealing, sexually spiced remarks, while Bachelorette No. 3 maintained her reserve.

Eventually, Cupid worked his work, and the sports anchor chose Bachelorette No. 3. To my relief, I had pegged the preferences of both TV anchors. In fact, I still wonder whether this second pair eventually fell in love. I say this because after the event I explained to Bachelorette No. 3 that her Builder date was likely to be a man of honor, duty and loyalty. Hearing me, he leaned toward her and whispered, "It's true."

Serotonin's Magic

Of the nearly forty thousand anonymous people I surveyed in my Personality Type Study, 28.6 percent were Builders. Of these, 27.6 percent were men and 29.4 percent were women, a slight but mathematically significant difference.

These men and women share many traits. Builders tend to be loyal and conscientious; duty, respectability and proper moral conduct are particularly important to them. Builders are also conventional; they admire and follow social norms and customs. They respect authority, follow rules and enjoy making plans and keeping schedules. They think concretely; Builders are often literal, detail-oriented and orderly, as well as cautious but not fearful. They tend to be social. And Builders are generally superb at managing people—at work, in the family and in their various social circles.

Although the sports television anchor may have known these things about himself, he undoubtedly did not know what produced these traits: serotonin, the mother of all neurotransmitters, the oldest of them all. Serotonin has extensive highways through the brain. You inherit your basic activities of this chemical. However, your experiences can alter this thermostat to some degree. And as this chemical floods, lingers and recedes from various brain regions, it affects your attitudes and behavior.[1]

No brain system acts alone, of course. Serotonin, for instance, can elevate estrogen levels in brain regions that influence thinking, emotion and memory. Serotonin can also trigger the release of oxytocin—a chemical associated with trusting and the ability to figure out others' states of mind. So all three of these chemicals probably contribute to the Builders' affability, their deep need to attach to family and community, and their tact and managerial skills.

Moreover, serotonin can suppress testosterone activity, particularly in a brain center associated with fear, anger and aggression—the amygdala. This may be why, in part, Builders tend to be calm and self-confident: their serotonin is suppressing aggressive tendencies.[2] Last, serotonin suppresses dopamine activity. And as you recall, dopamine often produces impulsive recklessness; hence it is likely that by reducing dopamine activity, serotonin generates the Builder's caution.

In short, serotonin has myriad interactions with other brain chemicals. Nevertheless, serotonin is central to the mind-set of the Builder.

Respect for Authority and Loyalty

A Builder par excellence was George Washington, the first president of the United States, a national symbol almost as powerful as the American flag. Washington had just the right mix of traits to lead the American rebels to victory, then fashion a new nation.

Foremost, Washington was a superb military man. As a young man he had served with distinction in the Virginia militia during the French and Indian War. Like the Builders in my Personality Type Study, he respected authority and was able to work effectively within a social hierarchy.

Washington was also loyal, another core trait of the Builder associated with serotonin. We all inherit some degree of loyalty, of course. And we learn the benefits of allegiance. But Explorers are devoted to their experiences. Directors are dedicated to their work. And Negotiators tend to be devoted to family, people and ideas. Builders, on the other hand, often feel a special obligation to the larger community. Builders see loyalty as duty.

Washington certainly did. Fighting at Lexington and Concord had produced "the shot heard round the world." And when Washington appeared at the meeting of the Second Continental Congress in May 1775, he was wearing his military uniform—a sign of his respect, allegiance to the American cause and willingness to serve in the Revolution. Washington did not wish to become commander in chief of the Continental Army. He was a modest man who had already resigned once from the military and was fully absorbed in developing his beloved estate, Mount Vernon. But Washington believed he had to contribute. And when his comrades elected him to lead American soldiers into battle, he felt it was his duty to accept this assignment.

Orderliness: The Rule Keepers

Washington was also orderly.

Builders pride themselves on being orderly—another trait associated with an active serotonin system. In my Personality Type Study, Builders responded far more enthusiastically than the other types to the statement "Taking care of my possessions is a high priority for me."

Builders are also orderly in their daily routines. Before Builders embark on a special job at work, they organize themselves, often breaking down the assignment into steps. They like to follow a prescribed process; they don't like unpredictability or working under pressure. Even on weekends, Builders are likely to make lists of what they want to do, as well as arrange their social engagements well in advance.

Builders also like rules, a trait that probably stems from their biological need for order. When asked to respond to the statement "In general, I think it is important to follow rules," Builders in my Personality Type Study were far more likely than other types to strongly agree.

Washington liked rules. Every morning of the war, after prayers, he would read the day's orders to his regiments. Sometimes he even authorized lashes for those who failed to act according to the rules. He even discharged some officers for straying off the path of good conduct, saying, "I spare none."

It's not surprising that Builders such as Washington (and many others) would gravitate to the military, or to government jobs, law enforcement, major corporations and rules committees. Builders feel comfortable in hierarchies—where duty and loyalty are required, and structure, rules and order reign.

The Plan Routine and Schedule Makers

A strong preference for careful planning is another trait of the Builder. In my Personality Type Study, Builders stood out in their agreement with the statement "I enjoy planning way ahead."

My friend Sarah is a consummate planner. Weeks in advance, she plans almost every hour of her next trip to London. She buys tickets online to musical events, makes restaurant reservations and knows the opening and closing hours of the museums she will attend. She says she likes making the plans as much as she likes following them. What would be cloying to an Explorer, impractical to a Director and regarded as inflexible to a Negotiator is relaxing to Builders like Sarah. Their logistical capabilities are supreme.

Builders also like schedules and routines. I learned this one evening some twenty-five years ago. That morning I made plans with Charles, a friend of mine, to amble from his apartment in New York's SoHo neighborhood to a tiny Spanish restaurant in the West Village, El Faro. It was a twenty-minute stroll. But as we prepared to depart, the sky turned black and it began to pour, a deluge that the local TV news station promised would last several hours. So I suggested we dine somewhere closer. His reply: "But we said we would go to El Faro."

"Habit is stronger than reason," wrote philosopher George Santayana. This need to stick to plans, methods and habits has been associated with serotonin and is a hallmark of the Builder. Only the Builders in my Personality Type Study heartily agreed with the statement "I think consistent routines keep life orderly and relaxing." In fact, because Builders like to meet their deadlines in an organized, structured and timely way, interruptions and diversions can irritate them. Their goal is closure; they want things to be resolved, settled.

Precision and Details: Step Thinking

His respect for authority, his loyalty to his cause, his taste for order and his need to follow rules and make plans and schedules probably all helped George Washington do his duty. But he shared another trait with most Builders: he was precise and detail-oriented. As one of his comrades in arms wrote of him, "Method and exactness are the *forte* of his character."

At Mount Vernon, for example, Washington meticulously arranged his flower garden and vegetable plots, his hedges and paths, his lawns and shrubs. Even the way he spaced his words on a page had to be perfect. And although he refused a salary for his job as commander in chief of the Continental Army, he promised to keep "an exact account" of his expenses.

To Builders, details matter. These people remember people's names, the birthdays of friends and relatives, details about neighbors and colleagues, and the exact date, time and location of specific family or community affairs. I suspect that many of the men and women who send "Christmas letters" that diligently recount the details of their year's doings are Builders. This precision is associated with high serotonin activity.

Builders must be organized and accurate, too. They are process-oriented. They like to pursue specific goals in a precise, consistent, sequential, linear, step-by-step way, what I call *step thinking*. In my Personality Type Study, only Builders heartily agreed with the statement "I like to work in a straightforward path toward completing the task."

Even in their everyday speech, Builders tend to impart their thoughts accurately, precisely and completely, providing all of the relevant information. If you interrupt them, finish their sentences for them or ask questions that appear superfluous, a Builder is likely to think you are not interested or not listening. To them, tangents and side conversations are frustrating distractions. Samuel Johnson once said, "A good conversation is a series of interruptions." Builders would not agree.

Their jokes also reflect their need for order, predictability, precision and closure. Builders don't tend to like incongruous humor, unresolved humor or nonsense humor. They prefer jokes and stories that are concrete and resolved—a taste in humor that has been linked to the activity of serotonin.

Persistence and Patience

Shortly after the War of Independence, Washington was overwhelmingly elected America's first president. He was dismayed, having long dreamed of returning to his beloved Mount Vernon to be a gentleman farmer. But, ever dutiful, he accepted the job in order to finish the work he had started: constructing a new country. Washington was persistent and patient—more Builder traits associated with serotonin activity.

I recently witnessed this persistence in a bookstore. I had purchased several books as birthday presents. The saleswoman wrapped my gifts with what I regarded as unnecessary precision. I was impatient to go, but she could not finish her task until every angle of the wrapping and every pattern in the paper were perfectly aligned. I even remarked to her, "You're so patient." She replied, "I was born that way."

"My strength lies solely in my tenacity," declared chemist Louis Pasteur. Builders are persistent and patient, indeed tenacious, in part because they can tolerate repetition. They are not easily bored, so they excel at tasks requiring sustained attention and repetition.

Interestingly, repetitive actions even raise levels of serotonin. This may be why some people repeat an action over and over when they are anxious, like checking to see if they have turned off the stove or rummaging through their suitcase to be sure they packed all they need for a vacation. As they repeat these motions, they could be unconsciously triggering serotonin activity in the brain, reducing anxiety.

Builders may also be relieving anxiety when they doodle. When members of Chemistry.com selected from an assortment of doodles the ones they were most likely to execute, the Builders chose grids and repetitive designs. This result supports research on serotonin showing that the Builder type gravitates to symmetrical designs, particularly pyramids, crosses and other geometric shapes that shout simplicity, order, predictability and repetition.

Remarkably, men and women who express serotonin activity (whom psychologist Marvin Zuckerman calls low sensation seekers) also have a taste for geometry in their clothing. As Zuckerman reported, "For some unfathomable reason low sensation-seeking males seemed to favor plaid sports shirts!"

Could today's golfers who wear plaid pants and hunters who don plaid shirts be unconsciously advertising their temperament with their clothing, in effect saying, "Trust me; I'm orderly, predictable and loyal; I'm a Builder"?

Fact-Oriented: The Number Crunchers

Builders focus on what they can see, hear, touch, taste and smell; they live in a universe of concrete reality. They are comfortable in clear-cut situations where the details are known and proven. Hence Builders collect facts, lots of facts. And, unlike the other types, they can remember these details without putting them into a broader contextual picture. So Builders are likely to be particularly good at games of trivia.

Not only do Builders like facts; they base their decisions on them. They tend to be quite literal, as well as suspicious of theories, inferences and conjecture, and uncomfortable with ambiguity. So if you want to impress a Builder on a date or in the office, don't give him or her your big ideas. Stick with details, precise information.

I should have remembered this advice when I was interviewed recently in Kansas City. The journalist had taken me to a funky coffeehouse near her newspaper's headquarters. We settled on a couch, and she began asking questions about the four personality types. Her queries were concrete and precise. That was fine. But I began to use the data to paint a larger picture, laced with theories. She became testy, which unnerved me. True to my Explorer self, I then became more and more theoretical—trying to win her over with what I regarded as exciting ideas. This proved to be a huge mistake, as I found out when I read her caustic review of my work a couple of weeks later. Builders want established facts; theories leave them cold.

Perhaps Builders' concrete, factual disposition contributes to their superb ability with numbers. Elevating the activity of serotonin in the brain actually inhibits general creativity and reduces the impulse to explore. But individuals with a specific genetic variant in the serotonin system have higher scores in "figural and numeric creativity." In fact, Builders handle detailed columns of figures with dispatch, in all likelihood because numbers are concrete, precise and orderly—the gifts of serotonin.

Managerial Genius

Not surprisingly, Builders make good managers and administrators. They follow the rules, stick to the facts and pay attention to the details; they are reliable, persistent and conscientious; and they reach out to those around them, building and maintaining their social ties.

This managerial savvy is yet another trait associated with serotonin. People who take serotonin boosters become more cooperative during group tasks. They make more suggestions, deliver fewer commands and make fewer attempts to solve problems by themselves.

George Washington was a meticulous plantation manager. He got up at dawn to ride around the perimeter of Mount Vernon, then often returned to read agricultural manuals, committing long sections to memory. He supervised his carpenters, blacksmiths, coopers, weavers, shoemakers, seamstresses, millers, fishermen and livestock. He even supervised the making of cider into applejack and the repairs on his freight sloop that plied the Potomac. And he wasted nothing.

As one of his biographers, James Flexner, wrote of him, "With his unbounded energy and his gift for detail, Washington was at Mount Vernon an efficient commander in chief."

Guardians of Tradition

Poet Rainer Maria Rilke called our attachment to traditions "the gnarled fidelity of an old habit." Builders like old habits: they are traditional. My Personality Type Study confirmed this. Only Builders heartily agreed with the statement "My friends and family would say I have traditional values." Builders also had a strong favorable response to the statement "Long established customs need to be respected and preserved."

Builders remember anniversaries. They enjoy business picnics, holiday parties and seasonal hunting, fishing, skiing or sun-worshipping vacations with friends. I suspect these men and women also predominate in summer colonies where old friends gather annually. "Historic continuity with the past is not a duty, it is only a necessity," wrote American jurist Oliver Wendell Holmes Jr. Builders agree; rituals, ceremonies and traditions appeal to them enormously.

Plato called this personality type the Guardians, the preservers of tradition, custom and community. And they certainly expressed their conservative values in my Personality Type Study. When asked whether they were "ultraconservative," "conservative," "liberal," "very liberal" or "other," Builders were significantly more politically conservative than the other types. Moreover, when asked where they wanted to live, Builders chose the suburbs or the country. No wonder so many midwestern Americans tend to be politically conservative; these states are full of Builders.

Scientists have long held that your political orientation comes, in large part, from your inherited biology. This, apparently, is true. Indeed this trait, observing social norms, has even been associated with specific genes in the serotonin system.[3]

Americans of all kinds enhance their affection for tradition during holidays by boosting levels of serotonin—with food. Take Thanksgiving. In the United States, most people celebrate Thanksgiving with a family dinner. And the core of the meal is usually turkey—a fowl loaded with L-tryptophan, an amino acid that is sucked into the bloodstream and hustled to the raphe nucleus in the brain where it is converted into serotonin. So as we dine with kin, we feast on food that makes us feel good about custom and tradition.

Rice, pasta, bread, cereal; under stress many turn to the heavy starches, "comfort food." People say these ordinary delights remind them of their childhood and home. They probably do, because these foods also boost serotonin—triggering feelings of calm, familiarity and tradition.

In fact, the ancient Greeks worshipped grain, through Demeter, their goddess of the harvest. Most of the Greek and Roman gods were capricious, jealous, quarrelsome or troublemakers. But Demeter was responsible, dependable, protective and giving. Her divine power, the Greeks believed, brought forth the fields of grain. So they worshipped Demeter in autumn—not with bloody sacrifices but with gentle prayers as they shared their first loaves of harvest bread, thereby elevating serotonin activity.

Cautious but Not Fearful

My friend Fred calls himself the "low-risk" type. We make quite a pair as we cross the streets in New York City. I don't actually weave around moving cars, but I don't necessarily stop walking at red lights either. Like many New Yorkers, I have developed a sense of how much time I have before becoming "roadkill." Fred stands rooted to the sidewalk until the light turns green. Indeed, Fred steps *back* at least two paces from the curb when the light turns yellow. Fred is prudent, another Builder trait.

In my Personality Type Study, Builders were far more likely to respond enthusiastically to the statement "I tend to be cautious, but not fearful."

This caution is to be expected. Several genes in the serotonin system have been associated with "harm avoidance." And men and women with specific combinations of these genes are known to be cautious but not fearful. Where dopamine triggers exploration, serotonin puts on the brakes.

One doesn't think of the commander in chief of the Continental Army, the man who led the United States to independence, as cautious. But Washington was. His first aide-de-camp, Joseph Reed, described Washington as he was trying to make a difficult decision: "[He] hesitated . . . more than I thought the public service permitted."

Prudent men and women are not necessarily averse to risk. They simply take particular comfort in the familiar. Builders seek security the way Explorers seek novelty, Directors seek achievement and Negotiators seek harmony. What better guardian could Americans have hoped for than the thoughtful, cautious, high-serotonin George Washington.

Conscientiousness

All these Builder traits add up to something special: conscientiousness. This splendid human attribute is a combination of several related characteristics.

Psychologist Marvin Zuckerman believes that conscientiousness is composed of two basic traits: dutifulness, or the degree to which people follow strict standards of conduct; and deliberation, the degree to which they plan and express caution and thoughtfulness. Other scholars have added more traits, regarding conscientious people as cautious, prepared, methodical, fastidious, compulsive, persistent, self-disciplined, thorough, orderly, dependable, dutiful, civic-minded and ethically principled. Conscientious people, they say, have a deep sense of duty. And they make plans, follow schedules and adhere to social norms.

The Myers-Briggs personality test adds that these men and women want to do things "in the accepted way" and "support the established methods." They prefer to "plan things far in advance" and "know ahead of time" what they will be doing. They like words like *scheduled, orderly, systematic* and *organized*. These men and women would rather *build* than *invent* and prefer to be *sensible* instead of *fascinating*. They seek certainty and order. They want to have *both feet on the ground*.

As I've mentioned, many of these subcategories of conscientiousness have been linked with specific serotonin activity. So it is not surprising that many Builders adopt a work ethic early in life. My niece, Piper, started working her newspaper route when she was seven—using her red wagon to cart the papers through the back streets of Aspen, Colorado, and hurling them at people's front doors.

Builders are conscientious and community-oriented. They often help around the house, volunteer at their place of worship or in their neighborhood, and do thankless tasks in the office, in their town or city, for their country or for the world—often with a selfless, stoical, charitable attitude. Washington vividly summed up the cost of this trait, too, saying to his staff, "Gentlemen, you will permit me to put on my spectacles, for I have not only grown grey, but almost blind in the service of my countrymen."

Builders serve.

Social, Calm and Beneficent

Builders are also social. In my Personality Type Study, Builders were significantly more likely than other types to describe their friends as a "social crowd."

Washington was endlessly social. He and his wife, Martha, entertained some two thousand guests between 1768 and 1775, harboring kin, friends, colleagues and just about any stranger who wandered past Mount Vernon at nightfall. If his dining room was empty of family, friends or passersby, Washington regarded it as "lonesome."

Not all Builders are social, of course. But certain genes in the serotonin system do contribute to sociability.[4] Men and women with these genes have more close friends. Even very shy people who take antidepressants that elevate serotonin activity in the brain and bloodstream become more gregarious.

Serotonin also makes you calm. This is a primary reason why people take serotonin boosters, such as Prozac and Paxil: to feel serene. Picture those popular boys or girls in high school who appeared so affable, calm and self-confident. They probably had particularly high activity in their serotonin systems.

Perhaps this desire for relaxed self-confidence is what attracts some people to psychedelic drugs. MDMA, known as "Ecstasy," as well as the related hallucinogens LSD, DMT and psilocybin, trigger a gigantic surge in the entire serotonin system, giving you feelings of intense well-being and comfort, as well as tactile sensitivity and bizarre perceptual effects. Ecstasy also releases dopamine and norepinephrine. But the serotonin surge produces the sociability and self-confidence, while dopamine triggers the euphoria.

Along with calm and friendliness, Builders often display beneficence. Washington again was a supreme example. In one of the great moments of American history, he turned down the offer to become America's first king. As Flexner wrote of Washington, "In all history few men who possessed unassailable power have used that power so gently and self-effacingly for what their best instincts told them was the welfare of their neighbors and all mankind."

High Status: Belonging

Washington would have been pleased to read Flexner's admiring words. Like many Builders, he was often worried about his reputation. To Builders, appearances matter. They take to heart these lines from William Shakespeare: "For the eye sees not itself / But by reflection."

Builders want to belong, to be regarded as reliable, respectable and charitable, to be members in good standing. They depend on their social networks for their emotional well-being. And they can be crushed by personal criticism. So they strive to maintain their ties, meet their responsibilities and behave with decorum. In fact, Builders are willing to endure tremendous personal sacrifice to remain embedded in the social hierarchy or even reach the top.

Their energies generally pay off, too. Individuals with specific genes in the serotonin system generally achieve higher socioeconomic status; they also tend to live in more affluent neighborhoods.

Moreover, Builders are proud of their accomplishments, displaying their medals, trophies, diplomas and certificates, and showing photos of themselves with family members, influential citizens or well-known politicians.

This correlation between serotonin and social success has been found in a wide range of other species. When given antidepressants that boost serotonin activity, vervet monkeys also climb the social ladder. Alpha males of the community have about twice the serotonin levels in their brain and blood as their subordinates. Even socially dominant lobsters have higher levels of serotonin.

So whether you're a human, monkey or lobster, if you have more serotonin activity, you are more likely to join, even reign in, the "establishment."

Religiosity and the Biology of "Ought"

You might expect that people who respect authority, follow the rules, believe in loyalty and tradition, and work in the "establishment" would also be religious.

Not necessarily. I know more than one atheistic Builder. My friend Bob, for example, is calm, social, popular, conscientious, superb with numbers and traditional in many ways, but a more ardent atheist you will never meet, even though he grew up as a Mormon.

Still, studies of identical and fraternal twins indicate that religiosity does have a genetic component. And people with serotonin activity in certain brain areas, including the raphe nucleus where serotonin is produced, tend to exhibit *self-transcendence* or religiosity.[5] Religiosity has also been associated with other aspects of the serotonin system. So variability in these serotonin pathways may explain why some people express more religious fervor than others. It also explains why Builders in my Personality Type Study were far more likely than other types to be part of an organized religion.

And no wonder psychedelic drugs—including Ecstasy, LSD, psilocybin and mescaline—often trigger feelings of spiritual awareness, mystical enlightenment and religious ecstasy. As I mentioned, these drugs dramatically elevate serotonin activity in many brain regions.

Probably related to Builders' religiosity is their need to do what is "right." Builders can be beleaguered by their conscience, what Darwin called "that short but imperious word, 'ought.'" *Ought; should; must; proper; right; wrong:* these words tend to alienate the freedom-loving Explorer, pragmatic Director and compassionate Negotiator. But Builders are comfortable with them.

Indeed, in my Word Type Study, Builders were particularly fond of the words *moral, morals* and *values.* And when asked to respond to the statement "People should behave in ways that are morally correct," only Builders in my Personality Type Study overwhelmingly agreed.

Oprah Winfrey would probably agree with that statement too. I suspect that Oprah is primarily a Negotiator. But she has some Builder in her, too, as the following remark suggests. When she announced her support for Barack Obama for president in 2008, she said, "I know him well enough to believe in his moral authority. And that is the number one reason why I am supporting him."

On the Dark Side:
Negative Traits of the Builder

"Everyone is a moon and has a dark side which he never shows to anybody," wrote Mark Twain. We often try to hide our weaknesses and quirks of personality, even from ourselves. But inevitably our faults reveal themselves. And just about every human asset can become a liability.

Builders, for example, often have a "right" way of doing things. But some believe their way is the only way, becoming dogmatic, closed-minded and stubborn. I know a young couple like this. Both are Builders. As she says, "We can't do household chores together. Last week we fought over how to lay down some carpet, and yesterday we had a raging battle over how to cook a turkey." Emotional

rigidity is linked with serotonin. And because Builders are committed to rules, schedules and traditions, they can bicker about a lot of
little daily issues.

Builders are realistic and cautious, often fine qualities. But these
traits can metamorphose into deep pessimism, a trait linked with
serotonin pathways. Their pessimism can turn to fatalism, too. Some
Builders, for example, maintain that nothing will ever change for the
better.

A related trait, perhaps, is the Builder's stoicism, also associated
with serotonin. This trait can be admirable, but it can evolve into one
of two extremes: self-loathing or feelings of moral superiority. The
Builder either fails to see any of his or her merits, or puffs and preens
with sanctimonious self-satisfaction while passing moral judgment on
others. Builders can be overly critical, too.

Builders are also frugal, yet another trait linked with serotonin.
In some, this can turn into stinginess. And because Builders are literal,
fact-oriented and have a "show me" attitude, they can fail to appreciate the emotional subtleties of a person or miss the nuances of a social, economic or political situation.

Builders can also become fixated on the past. As a Builder girlfriend
of mine said recently, "I can daydream for hours about what life would
have been like if I had bought another house instead of the one I bought.
Or what life would be like if I had married my old boyfriend instead of
my current husband. Or how I could have gotten better grades in school
or made more money. I really live my life with regret." If Explorers live
for today, and Negotiators live for tomorrow, Builders can wallow in
their thoughts of yesterday.

Perhaps most dangerous is the Builder's loyalty and respect for
authority. These traits are often admirable, but not always. Take General Colin Powell of the United States Army. Powell is a Builder; he is
known worldwide for his competence, reliability, decency and loyalty.
Yet Powell's sense of loyalty may have helped lead the United States
into war with Iraq—despite his conviction that this was monumen-
tally unwise. Many people believe he tipped America toward war

when, on February 5, 2003, he appeared before the United Nations and maintained that the Iraqi dictator Saddam Hussein possessed weapons of mass destruction. Before his speech, two-thirds of Americans were against a war; afterward, half were in favor of invasion.

Powell had been deceived into believing that Hussein was building bombs. But he *chose* to support President George W. Bush's solution: attack. In a similar situation in the past, he had explained his actions by saying, "I had done what my conscience and duty required." Due, I believe, to his disposition to respect authority and follow orders, Colin Powell contributed to chaos in the Middle East.

Possible Physical Problems

Scientists report that specific genes that increase activity in parts of the serotonin system also inhibit "task switching." In other words, these people have difficulty stopping what they are doing. So perhaps the serotonin system plays a role in a peculiar behavior pattern, "hoarding." I have a physician friend who hoarded for years. Her house was so filled with "stuff" that she could no longer sleep in her bedroom or eat at her dining room table—until she began to take an antidepressant that triggers dopamine activity, thereby suppressing serotonin.

Hoarding is a complex addiction. But I suspect it may stem, at least in part, from the biological font of persistence: serotonin.

Builders: The Pillars of Society

American financier Robert Morris wrote to George Washington, "Heaven . . . has blessed you with a firmness of mind, steadiness of countenance, and patience in suffering that give you infinite advantages over other men."

Builders, like Washington, have many strengths. They tend to be calm, modest and self-effacing, as well as affable and socially skilled. Builders are hardheaded; they have a clear memory for yesterday's mistakes, so they prepare. They are steady, reliable, cooperative, in-dustrious, hardworking and patient. They are dedicated to family and community. And Builders work diligently to preserve traditions and standards. They are often fine protectors and providers.

Thanks to the persistence of human biology, little has changed since Plato called these men and women Guardians some 2,500 years ago. Builders are still the "pillars of society."

CHAPTER 6

Only he who keeps his eye fixed on the far
horizon will find his right road.

—DAG HAMMARSKJÖLD

ALWAYS TO THE STARS:

The Director

In a jar, in a cider box, under the beer cooler in his office, pathologist Thomas Harvey tucked Albert Einstein's brain. Harvey had begun to perform scientific studies on this brain soon after Einstein died in 1955, taking photos, measuring it and cutting it into 240 pieces. But he published nothing. Finally, in 1996, Harvey sent a one-sentence fax to neuroscientist Sandra Witelson at McMaster University in Ontario: would she like to study Einstein's brain?

Soon Witelson began to compare 50 pieces of Einstein's brain with those of 91 other deceased men and women, including singers, architects and blue-collar workers. To her astonishment, she found that Einstein's inferior parietal lobes (regions behind the ears where we process mathematical problems and visualize three-dimensional spatial relations) were some 15 percent wider than normal. Moreover, the deep groove that runs through this region, the Sylvian fissure, was largely pushed aside.

This may explain why Einstein's mother had worried that her infant son was retarded. At birth his head was enlarged and oddly shaped on the sides. But this unusual brain architecture must have enabled nerve cells in these brain regions to work together more efficiently. Indeed, with his unique mental equipment, Einstein would change our understanding of the universe.

Albert Einstein was a classic Director. These men and women speak their mind. They are direct, tough-minded and decisive. Yet before they make a decision, they examine their choices thoroughly and unemotionally. Directors admire self-control; they are analytical, skeptical and exacting. Directors are independent, too. And they must achieve. Indeed, they enjoy competing to get to the top. Therefore, many are pragmatic, focused and daring. Directors can also be inventive, as well as mechanically or mathematically skilled. Oddly, many are musical, particularly adept at understanding the structure of music. Others are highly skilled at spatial games, from football to chess. And many Directors have a heroic side; they are the men and women who dash into a burning building to save a stranger.

Directors composed 16.3 percent of the sample in my Personality Type Study. Of these Directors, 24.8 percent were men and 9.7 percent were women—a dramatic difference. I was pleased with this breakdown by gender; it indicates that the traits of the Director are associated with the predominantly male hormone, testosterone. Both sexes produce testosterone, but most men produce far more.

I know many female Directors, of course, among them, Cathy. Cathy has smooth pink-white skin, huge blue eyes, a bouncy blond ponytail, a high, soft voice and all the other appealing traits of a vibrant, seductive young woman. But beneath her beguiling exterior, Cathy has the keen mind, ambitious spirit and exacting attitude of a true Director. Most important, like Einstein and other Directors, Cathy excels at what scientists call *systemizing*.

Systemizing

Psychologist Simon Baron-Cohen distinguishes two gender differences in patterns of thinking, which he refers to as *systemizing* and *empathizing*.

Systemizing is the propensity to construct and analyze systems, from building bridges or fixing motorcycles to studying the ecology of a pond. Empathizing is the ability to identify with and respond appropriately to another's thoughts and feelings; empathetic people are intuitive and compassionate. Baron-Cohen further maintains that systemizers express more testosterone and are largely men, while empathizers express more estrogen and are more often women. But as he says, some men and women are equally proficient at both empathizing and systemizing; and some women excel at systemizing, while some men excel at empathizing.

This is exactly what I found when analyzing the results of my Personality Type Study. When responding to the statement "I understand complex machines easily," Directors agreed far more than other types. These people also responded dramatically to the statements "I like to figure out how things work" and "I am intrigued by rules and patterns that govern systems." More Directors were systemizers, and more Directors were men. Nevertheless, whether it's understanding computers, automobiles or electronics; the habits of wolves; logic, math or music; voting patterns; or the physics of your tennis swing: if you are skilled at understanding rule-based systems, you have a core trait of the Director.

Einstein was a supreme systemizer. His mission to dissect the systems that govern the universe first emerged in 1895, when, at age sixteen, he began to think about riding beside a beam of light. As he later wrote, "I became more and more convinced that nature could be understood as a relatively simple mathematical structure."

Testosterone Priming: Finger Talk

Systemizing is linked with testosterone, and it's a skill you begin to acquire before you leave the womb.

As testosterone washes over the fetal brain, it builds the brain architecture for this systemizing ability. Meanwhile, surges of fetal estrogen begin to construct the brain anatomy for empathizing. Levels of both fetal testosterone and fetal estrogen swell and recede in patterns that are inherited. So, depending on the hand you're dealt, you will express predominantly testosterone, predominantly estrogen, or a lot or a little of both—thus predisposing you to be better at systemizing, at empathizing, or good or bad at both.[1]

Testosterone leaves its mark in many ways. Take a look at your right hand, palm up. The more testosterone you were exposed to in the womb, the longer your fourth or ring finger will be in relation to your second or pointing finger. If you were exposed to more estrogen, these two fingers will be the same length, or your pointing finger will be longer than your ring finger. Once again, this is exactly what my Personality Type Study showed: Directors tended to have a longer ring finger, while Negotiators had a ring finger that was the same length as or shorter than their index finger.

Bill Gates must be a Director. I recently watched him give a speech at a conference. His image was broadcast on a huge screen in the center of the stage, and periodically he held up his hand to accentuate a point. Everyone else in the audience was focused on his words. I was fixated on his right hand. Finally, he held it high and steady just long enough. Indeed, his ring finger was considerably longer than his pointing finger—as one would expect of a world-class systemizer.

Not all adult Directors receive this surge of testosterone in the womb. So if you are a Director and don't have a longer fourth finger, it simply means that you were flooded with testosterone at a later time in life, most likely at puberty.[2] One way or another, testosterone primes Directors to systemize, as well as excel at many other things.

Spatial Skills

According to his sister, Maja, Einstein could construct a fourteen-story house of cards while still a toddler. Testosterone has long been associated with superb spatial skills in both sexes—although most of the studies have been done on males. Nevertheless, these data provide some fascinating insights into this trait of Directors, both men and women.

Most one-day-old boys look longer than girls at mechanical and moving objects. Infant boys are better than most infant girls at tracking a blinking light across a TV monitor. Boys like toy cars, blocks, trains and mechanical things they can construct or move around. Boys pack the video arcades to zap electronic enemies in spatial games. And when testosterone inundates the male brain at puberty, boys in many cultures begin to outstrip most girls in math, geometry, mechanical drawing, mechanical reasoning and engineering.[3] Data from my Personality Type Study show that far more Directors are physicians, mechanics or engineers.

Fetal testosterone builds this spatial prowess in an ingenious way. Spatial talents lie largely in the right hemisphere of the brain. As fetal testosterone surges, it builds the right hemisphere faster than the left, compromising the left hemisphere and endowing the individual with superior spatial skills.

Keep in mind, though, that bodily levels of testosterone can be enhanced by puberty or other experiences, thus contributing to your spatial dexterity. For example, women are better at finding their cars in a parking lot around menstruation when estrogen levels decline, enabling testosterone to express itself.

The correlation between testosterone and spatial aptitude is not simple. Too much or too little of this powerful chemical can dampen your spatial prowess. And your daily habits and environment always play a role. But the link between spatial skills and testosterone is strong.

Musical and Athletic Ability

This spatial aptitude gives many Directors an unexpected gift: musical talent.

I love music of many kinds, but when I listen, I just swing to the beat. Directors also notice the composition of the music, the structural arrangement of the notes. And some Directors are brilliant at composing and playing it as well. Central to understanding advanced musical composition is spatial talent. And testosterone priming in the womb contributes to this aptitude.

How curious. Why would fetal testosterone produce musical ability? Darwin puzzled over the evolutionary origins and purpose of music. He finally concluded that, like birdsong, "musical tones and rhythm were used by our half-human ancestors during the season of courtship."

Current data support Darwin's hypothesis. Men are ten times more likely than women to produce a jazz, rock or classical music album. And they produce their music at the average age of thirty—the height of a man's reproductive years. Most telling, male musicians tend to have longer ring fingers, which is associated not only with elevated levels of fetal testosterone but also with a higher sperm count. Millions of years ago Director types must have advertised their reproductive vigor with song.

Men who received more testosterone in the womb also excel at a number of sports, particularly sports that require spatial skills, such as soccer. And women around the world flock to admire these high-testosterone athletic men.

But a lot of Director men and women use their musical or athletic talent for other purposes. Einstein used his musical ability to think. His mother gave him a violin when he was five, and he loved music all his life, particularly pieces by Mozart. Neighbors often heard him playing his violin in his kitchen late at night as he struggled with some aspect of his theory of general relativity. One recalled, "Then, suddenly, in the middle of playing, he would announce excitedly, 'I've got it!'"

Focus

"Chance favors the prepared mind," wrote Louis Pasteur. Directors believe it. Directors are goal-oriented, so they tend to focus intensely as they work. But unlike Explorers, who focus on a wide range of interests, Directors concentrate narrowly and deeply—with the help of testosterone.

Fetal testosterone compartmentalizes the brain. In effect, it builds more short nerve cables that connect nearby parts, but fewer long-range cables connecting distant brain regions. This brain wiring enables the systemizing Director to focus narrowly and thus pay careful attention to the system's parts without being distracted by unnecessary data.

Testosterone injections in adulthood can also affect one's focus, as transgenders report. After three months of testosterone shots for a sex change from female to male, one individual said she became "more blinkered" and changed "in mental focus from broad to narrow." Perhaps this is why Directors tend to be nitpickers and hairsplitters. They are often focusing on a system's parts rather than on the whole. Perhaps this focus also drives Directors to concoct and relish puns and paradoxes. These groan-producing forms of humor structurally juxtapose disparate ideas.

The Director's focus can be wonderful, especially when it is centered on you. But it can be unsettling, too. Just ask a Director who is reading or watching television a simple question. The Director probably won't hear or see you. In fact, I have wondered if a Director in my life would notice if I rode past him through his living room on a rhinoceros or started eating pages out of his dictionary while he read nearby.

Analytical, Logical and Direct

Directors pride themselves on being logical. In my Personality Type Study, only Directors overwhelmingly agreed with the statement "I am more analytical and logical than most people."

Einstein would have been among them. "As a boy of twelve," he later recalled, "I was thrilled to see that it was possible to find out truth by reasoning alone, without the help of any outside experience." Indeed, Einstein worked in later years to develop his own son's analytical skills. As he wrote, "I want mainly to teach him to think, judge and appreciate things objectively."

A spinoff of the Director's unusually logical and focused mind seems to be a penchant to be outspoken and blunt—as I discovered in my Personality Type Study. Only Directors were highly enthusiastic about the statement "I like to avoid the nuances and say exactly what I mean." They also stood out in their positive response to the statement "I think it is important to be direct." Indeed, men and women who express high testosterone activity are less likely to strive to be polite, respectful, considerate or friendly. As Einstein once said, he had a hard time keeping "my big mouth shut."

Like Einstein, many Directors are abysmal at meandering social chitchat. Wasting words, repeating the obvious—it all seems pointless to them. Directors like to dispense with what they regard as extraneous facts and focus on the goal. These are the people who sputter, "Get to the point."

This economy of words can be both troubling and reassuring. A friend of mine recently asked her Director husband why he rarely said "I love you." He replied, "Sweetheart, I said that last month and nothing's changed."

Tough-Minded, Decisive, Exacting and Skeptical

In conjunction with their analytical, focused and direct approach, Directors are pragmatic. Respect for hallowed ways and social or political correctness irritate Directors. If they are forced to observe a tradition that makes no sense to them, they do it—grudgingly. Where Explorers are generous, Builders stoical and Negotiators altruistic, Directors are practical.

These paragons of expediency are also decisive, a trait associated directly with testosterone—as one sex-change patient noticed. After three months of testosterone injections to convert himself from a woman into a man, he mused, "I act faster." In my Personality Type Study, Directors stood out from the other types when they agreed with the statement "I have no trouble making a choice, even when several alternatives seem equally good at first."

Builders are slow to make important decisions because they are cautious and meticulous. Negotiators tend to ruminate because they see so many alternative solutions and want to imagine all of them. Explorers are somewhat faster in making decisions because they are impulsive. But my Personality Type Study showed that Directors are far more likely to make decisions quickly, even more likely than Explorers—perhaps because Directors are also so tough-minded. Only Directors strongly agreed with the statement "I am tough-minded."

"Action is eloquence," Shakespeare wrote. Directors agree. I know a Director who bought an expensive car in twenty minutes—and he wasn't rich.

Yet Directors are skeptical, and this distrusting nature is associated with a strong derivative of testosterone, dihydrotestosterone. Directors seem to follow the advice of the English prelate Thomas Fuller, "Trust thyself only, and another shall not betray thee." The only things that Directors truly believe in are reasoning, logic and systematic investigation. Directors are suspicious of facts until they have personally checked them out; as for motives, these they must carefully analyze.

The epitome of the skeptical Director was Metternich, the nineteenth-century Austrian statesman. When he was informed that the czar of Russia had died, he is said to have muttered, "What could he have meant by that?"

Resourceful, Self-confident and Bold

Directors can be ingenious at solving problems. Resourcefulness is as important to them as adventure is to the Explorer, loyalty is to the Builder and empathy is to the Negotiator.

Einstein certainly was resourceful in getting his first wife, Mileva
Marić, to divorce him. He had fallen wildly in love with this dark-
haired Serbian math student soon after he met her in his college
physics class. First they had a daughter; then they married and had
two sons. Deeply they loved; intensely they discussed ideas; meticulously she checked the math and logic in his school papers. But with
time they grew apart. And to extricate himself from their marriage,
Einstein made a bargain with her. If she would give him a divorce, he
would someday win a Nobel Prize and give her the prize money.

Seventeen years later she collected.

Einstein was self-confident and bold, like many Directors. As
long as Directors know the essential facts and are satisfied with the
reasoning behind the plan, they feel secure. Then they act, boldly.

Another bold Director was Martin Luther, the father of the German Protestant Reformation in the sixteenth century, who publicly at-
tacked the corruption in the Roman Catholic Church. Luther defiantly
stated, "Here I stand. I cannot do otherwise. God help me. Amen."
Like Einstein, Luther would change the course of history—with self-
confidence and boldness.

Confidence and daring: these traits arise, at least in part, from the
activities of testosterone. Menopausal women provide good examples.
With menopause, levels of estrogen decline—unmasking natural levels
of testosterone. And around the world, middle-aged women become
more action-oriented, confident, forthright and bold. All these traits
have been associated with testosterone.

An old school buddy of mine must be a Director. As founding
president of a merchant bank, he lends money to other banks, as well
as to enormously wealthy businessmen. One evening he went to a
fancy dinner in New York to negotiate the sale of a sizable piece of
New York real estate in the heart of midtown. By dessert he had completed the deal with the man sitting on his left. Instantly he felt bored.
So in a "New York minute," as he told me, he turned to the man on
his right and started his next big deal. He was smooth, persuasive and
bold.

Autonomous and Competitive

"I am the master of my fate: I am the captain of my soul." Poet William Ernest Henley expressed another facet of the Director: a fiercely independent spirit.

Directors can tolerate extreme isolation, long hours at the desk and many other discomforts when they work. But they must pursue their goals and solve their problems on their own. Directors demand autonomy—another trait tied to testosterone priming in the womb.

Einstein was highly independent. In a letter to a friend, he wrote, "A foolish faith in authority is the worst enemy of truth." And to a lover he remarked, "Long live impudence! It is my guardian angel in this world."

Directors have almost no respect for diplomas, references or credentials, either. Perhaps this is why Bill Gates dropped out of Harvard in 1975 during his junior year. In his mind, he had something better to do than collect a bachelor's degree from this revered institution: design innovative software and launch Microsoft. Like Einstein, Gates despises anything that smacks of authority or rules. Gates has worked for himself all his life. His independent approach has worked, too. Gates became a multimillionaire at age thirty-one; he was knighted by Queen Elizabeth at age fifty; and he has transformed modern society with his systemizing—and independent—brain.

Coupled with this hunger for autonomy and independence, the Director is competitive and aggressive. Explorers, Builders, Directors and Negotiators probably exhibit no real difference in *internal competitiveness,* the desire to meet personal goals and display excellence. But high-testosterone men and women score higher in *external competitiveness,* the willingness to elbow others aside to win.

In my Personality Type Study, Directors were distinctly more likely than other types to strongly agree with the statement "I enjoy competitive conversations." They were also more positive in their agreement with the statement "Debating is a good way to match my wits with others."

A journalist offered me a fine example of this trait one evening after I had made a speech. He came up to the podium and told me that he and his wife played golf. They often played with three other couples in a jovial, relaxed manner. But one morning they switched the ground rules: the four women played together, as did the men. After the match, his wife recounted how she and her girlfriends had skipped the fourteenth hole, and how one woman moved her ball out of the rough to take a better shot. "No one cared," she said. Her journalist husband wagged his head as he told me about his morning with "the boys." Their game had turned into cutthroat golf.

The association between testosterone and aggressiveness is well established. For centuries farmers have castrated bulls and roosters to curb their aggressive nature. And reams of data link aggressiveness with testosterone in both men and women.

Aggressiveness has long been celebrated, too. The classical Greeks believed that Ares, the god of war, gave men the ferocity to kill lions, slay dragons, fight enemies and protect their homes. And the Romans dressed Mars in shining armor, invincible and magnificent. The message: aggression wins.

But why must Directors "win" with such determination? If you ask them, they are likely to give you a host of practical reasons. But from the evolutionary perspective, this competitive striving gives Directors something they seek: rank.

"Big Men": Dominance Matching

Around the world, men and women who are exposed to more testosterone, either in the womb or as they age, tend to compete more aggressively for rank. This is particularly evident in the office.

During business meetings high-testosterone men are likely to break a silence to seize the floor, then deliver a "take charge" speech. They often attack with words. And they take coworkers more seriously when they argue back. As a colleague speaks louder, they do, too, triggering a spiral of *dominance matching* that sometimes ends in a shouting match.

I experienced this phenomenon one evening in New York. I was attending an event and standing with four men, one of whom began to bludgeon me with words. He was enjoying his snide remarks. The other three looked on as I curled my shoulders, hung my head and smiled deferentially in hopes of stopping the onslaught. But he was unstoppable—on he railed. Suddenly something snapped in my head, and I heard myself deliver a vicious retort. I have forgotten what it was, but I remember that it was not only stinging but funny.

Immediately all five of us stiffened and grinned, lips drawn back, teeth clenched in the primate "nervous social smile." A sickening silence ensued. And as I stood there waiting for his reply, several thoughts tore through my mind. Would he hate me forever for embarrassing him in front of his friends? Or would he respect me for matching his attempt at dominance? Suddenly he threw his arms up, laughed and hugged me. He has liked me ever since.

Nature isn't tidy. There is no simple correlation between testosterone and status. Too much testosterone, for example, is often associated with *low* rank. Moreover, this hormone never works alone. As mentioned earlier, Builders who have specific activity in certain serotonin pathways achieve higher ranks. And serotonin and testosterone often suppress each other. So it is not surprising that high-ranking professional men tend to have somewhat lower levels of testosterone than do blue-collar and unemployed men. In short, bodily testosterone needs to be at a specific level in relation to various other chemicals to correlate with rank.

Nevertheless, high-testosterone men tend to strive for rank. Why? Because women like high-ranking men, or what anthropologists call *big men*. Zulus; Aleut Eskimos; Mbuti pygmies; Americans: women everywhere would like to have a man's resources as they rear their young. Even female mice, deer and bighorn sheep find high-ranking males more attractive. So high-ranking men win sexual and reproductive opportunities—and more children, the evolutionary payoff for their competitive spirit.

Women who experienced high levels of testosterone in the womb are more likely to pursue male-dominated occupations and achieve higher-status jobs. But they are less likely to marry, and they have fewer children. So why, from the evolutionary perspective, do they strive to get ahead? Perhaps their ambitious nature evolved, in part, to help their biological relatives survive, thereby unconsciously passing on some of their own DNA.

Achievement and Knowledge

Regardless of the evolutionary payoffs, Directors yearn to succeed. They work hard and long, and they set high standards for themselves. As Winston Churchill put it, "I am easily satisfied with the very best."

But Directors rarely feel they have expended enough time and energy on their projects. And because many set their standards higher and higher, they become more and more tense—worrying that "this time" they will surely fail. Some Directors live with an almost constant sense of impending doom.

This need to achieve even permeates their leisure hours. They want to lift heavier weights, jog faster or play a better hand of poker. And when they disappoint themselves, they can be very self-critical. One of my friends recounted to me how she, as a child, used to hear her father as he sat in the bathtub muttering to himself, "Daddy stupid; Daddy stupid; Daddy stupid." Directors can be ruthless on themselves.

Yet they never give up striving, and many strive for one of life's highest prizes: knowledge. In my Personality Type Study, only Directors strongly agreed with the statement "I pursue intellectual topics thoroughly and regularly." Like Explorers, Directors need to know. But unlike Explorers, who have a wide range of interests, Directors focus their knowledge. In my Personality Type Study, Directors were the most likely to have a doctoral degree.

And Directors spend their spare time with those who share their thirst for knowledge. Only Directors reported that they socialized largely with "intellectuals." Moreover, in my Word Type Study, the words Directors used most often when describing what they were looking for in a partner were *intelligent* and *intelligence*.

A Few Great Friends

One thing Directors don't seek, however, is a wide circle of friends. Einstein was always a loner, even as a child. Every Thursday for several years, for example, his parents invited a poor student, Max Talmud, to share their dinner. This weekly event began when Max was twenty-one and Albert was ten. Years later, Max recalled, "In all those years, I never saw him reading any light literature. Nor did I ever see him in the company of schoolmates or other boys his age."

Testosterone correlates negatively with socialization. Infants exposed to higher levels of testosterone in the womb make less eye contact and have a smaller vocabulary; by age four they also show less social savvy and begin to develop fewer—yet deeper—interests. Be it the naval battles of World War II, the life of Michelangelo, the evolution of the mammalian or the latest strategies in chess, Directors need to know every last detail about their particular field of interest.

And they don't suffer fools gladly. Directors want to have intense discussions with a few close friends or people with whom they can share ideas. What did young Einstein talk about at the dinner table with Max Talmud? Science.

Nevertheless, Directors can be charming—if they are interested in the conversation. Negotiators are particularly fond of the forthright and knowledge-packed conversations they can have with Directors. I certainly am; like many Negotiators (my secondary type), I tend to see too many hidden meanings, innuendos and complexities in almost any social exchange, so I find Directors relaxing. I feel I can count on them to say what they mean. Moreover, they don't tend to exaggerate, like the flamboyant Explorer. They don't judge, like the moralistic Builder.

And they aren't manipulative, like the subtle Negotiator. Most important to me, what Directors talk about can be richly informative.

But if you want to communicate with Directors, ask them what they think, not how they feel. Directors don't talk easily about their feelings.

Emotional Containment: The "Right Stuff"

"If you can keep your head when all about you / Are losing theirs . . ." So begins Rudyard Kipling's famous poem "If." Directors fear losing their heads; they admire emotional control. To them, objectivity is essential. They feel it is imperative to be detached from their feelings, particularly when they make decisions. So Directors aspire to be calm, cool and collected. In fact, Einstein was hailed by journalists as the "embodiment of pure intellect" because his reasoning so completely trumped his inner emotional world.

Some of the men and women who have taken the Myers-Briggs personality test have shown this trait in spades. These individuals, whom Myers called the Thinking type, "value logic more than sentiment." They seek to be "just," "objective," "analytical" and "convincing." They believe it is a higher compliment to be called "a consistently reasonable person" than "a person of real feeling." They would rather be regarded as "competent" than "compassionate." And when making a decision, these men and women believe it is more important to "weigh the facts" than to "consider people's feelings and opinions."

The Directors in my Personality Type Study shared this perspective. When asked to respond to the statement "I am able to solve problems without letting emotion get in the way,"

Directors were far more likely than other types to strongly agree. Directors also believe, as Goethe wrote, "it is in self limitation that a master first shows himself." Unlike the Explorer, who can enjoy being out of control, the Director regards loss of self-control as weakness.

Testosterone is associated with this emotional containment. As levels of testosterone increase in teenage boys, many begin to mask their feelings of vulnerability, loneliness, anxiety and guilt. High-testosterone men and women smile less and employ less eye contact. Those with the "right stuff," as author Tom Wolfe called it, even develop the ability to contain feelings of intense fear and severe pain.

Heroic Altruism

Perhaps it is, in part, this self-containment that enables Directors to express an outstanding human trait: heroic altruism. Individuals with high levels of testosterone are more likely than other types to dash into a burning building to save a stranger, attack an armed bully with nothing but their fists, or brave a hurricane or tornado to save an abandoned dog.

And when asked what prompted them to perform their act of courage, these heroes often say they were barely thinking. "I did what I had to do," they say, or "It wasn't a conscious choice." These Directors are being honest, as Directors often are. Their brains are most likely flooded with testosterone—giving them strength, focus, decisiveness and resolve, driving them to act quickly and boldly.

Other personality types can be altruistic, too, of course. I suspect many Explorers share this heroic impulse because they are so spontaneous and energetic. Builders are cautious, however. Some can be extremely brave, but I suspect these men and women are less likely to run into a burning building without thinking through the consequences step by step. Negotiators are compassionate; many work long and hard to contribute to world peace, educate the poor or perform other altruistic acts. But with their contextual view of life, Negotiators are more likely to stand outside the burning building weighing the alternatives of risking their lives to help a stranger versus their responsibilities to their own children, family and community.

I suspect that most of the firefighters, police officers and other men and women who put themselves "in harm's way" every day are high-testosterone Directors.

On the Dark Side:
Negative Traits of the Director

"The keenest sorrow is to recognize ourselves as the sole cause of all our adversities," wrote Sophocles. Indeed, we bring many of our troubles upon ourselves.

Take the Director's emotional containment. There is a flip side to this propensity: emotional flooding. Despite their earnest efforts at self-control, high-testosterone men and women tend to have trouble containing their emotions. Directors are particularly susceptible to anger, exploding into a generalized rage when they feel hurt, disappointed or frustrated. In short, they "lose it." A pounding heart; tense muscles; sweating; shallow breathing; skyrocketing blood pressure; a burst of adrenaline and the stress hormones—the Director's entire body gets ready to fight or flee. Many lash out, too: scowling, screaming, even hitting. They can become swamped by what poet T. S. Eliot called "undisciplined squads of emotion."

Einstein suffered from emotional flooding, particularly during childhood. His sister, Maja, described his fury: "At such moments his face would turn completely yellow, the tip of his nose snow-white, and he was no longer in control of himself." Once he threw a chair at a tutor—who fled and never returned. He must have hurled more than a few household items at his adoring sister, too, because Maja later wrote, "It takes a sound skull to be the sister of an intellectual."

Directors are forthright, a fine trait when it's appropriate. But they can become blunt and abrasive. Because they are exacting, Directors can become uncompromising and demanding. Their ambition can lead them into workaholism. With their narrow focus, some Directors can miss the broader implications of their actions and behavior, trampling others. Their need for rank and recognition turns some into aggressive bullies and powermongers. And because Directors value autonomy so highly, they can shun emotional connections with those who love them.

Possible Physical Problems

Directors may be more susceptible to certain ailments and diseases, too. Fetal testosterone can *over*prime the brain. And as this chemical enhances growth in the brain's right hemisphere to produce superb spatial abilities, it can retard growth in some of the language centers in the left hemisphere—predisposing Directors to dyslexia, stuttering and other tics of language fluency. Einstein, for example, didn't speak until almost age three. As his sister said, "He had such difficulty with language that those around him feared he would never learn."

Individuals with a great deal of fetal testosterone also tend to be less skilled at recognizing emotions and less sensitive to the feelings of others. Affiliated with *mind blindness,* these people aren't able to judge what others are thinking or feeling. They can lack empathy, too, and become aloof and impersonal, even cold. In fact, Galen called these men and women "phlegmatic," largely because they were often distant and detached.

Moreover, scientists believe that an extreme overload of prenatal testosterone can contribute to autism and related syndromes. Severely autistic boys and girls make little or no eye contact with others, nor do they express compassion or even the slightest awareness of the feelings of others. Yet autistic individuals sometimes have a special and dramatically developed systemizing skill, such as knowing in an instant what day of the week it was on May 17, 1184.

Shooting for the Stars

Despite these possible problems, most Directors remain healthy and have many wonderful traits. With their gift for logic and reasoning, they can ignore the quotidian demands of life and concentrate on its challenges and mysteries. And with their bold spirit and penetrating focus, Directors can become intellectual geniuses, fine sportsmen and exciting friends and lovers. As Einstein said on his deathbed, "It is time to go. I will do it elegantly." He had high standards to the end.

Semper ad astra: always to the stars. I have often gazed at this inscription on my mother's family crest and wondered who chose this motto in sixteenth-century Holland. It was surely a Director, for it captures the heart and soul of these men and women.

Directors shoot for the stars.

CHAPTER 7

I remember everything simultaneously;
Like the distant beam of a distant lighthouse,
I carry the universe before me
Like an easy burden in an outstretched palm,
And in the depths, mysteriously growing, is the seed
Of what is to come.

—ANNA AKHMATOVA

THE PHILOSOPHER KING:

The Negotiator

"Nature is usually wrong," declared painter James McNeill Whistler. As a young man, Charles Darwin might have agreed; he was annoyed by much of what he saw in the natural world. Crimson snouts; chartreuse feathers; spots and stripes and dapples; the cock's comb and lion's mane: these accessories, he reasoned, were useless for survival in nature, red in tooth and claw. Darwin was particularly irritated by the peacock's tail. As he wrote to his son, "The sight of a feather in a peacock's tail, whenever I gaze on it, makes me sick." These flamboyant, cumbersome, seemingly worthless feathers undermined his theory that all traits evolved through natural selection—for a purpose.

Ultimately, though, this frustrating puzzle galvanized him to propose a second theory: sexual selection. Women's beauty, men's beards and many other natural decorations on all sorts of creatures evolved not to survive another day but to attract the opposite sex or fight contenders—to win the mating game.

Even as a child, Darwin was interested in nature. His brother called him "Gas" because of his many disastrous chemistry experiments. But he spent his happiest hours roaming the woods along the river Severn near his home in England's countryside. Here he collected bird's eggs, shells, beetles, fish, frogs, minerals and fossils. As he said, "I was a born naturalist." His kind but stern father regarded him as a something of a failure, however, saying, "You care for nothing but shooting, dogs, and rat-catching. You will be a disgrace to yourself and all your family."

At the time, Darwin had, indeed, failed to become a doctor, then a clergyman. He wasn't interested. But when, at age twenty-two, he got an opportunity to be a naturalist on board the HMS *Beagle,* which was heading for a surveying mission off the coast of South America and then around the world, he leaped for it. Despite his family's misgivings, Darwin set sail on this "floating coffin," as his father called it, on December 27, 1831.

Five years later he returned—and began to sort through the endless boxes of his collected rocks, exotic plants, birds and mammals. Darwin would write seventeen books and more than one hundred scientific papers on phenomena as different as orchids, barnacles and earthworms. But it was his grand *synthesizing* theories of natural selection and sexual selection that would explain the origins of and connections between all living creatures. Biologist Richard Dawkins would call this set of principles "the most important idea to occur to a human mind."

Darwin was a Negotiator. He was imaginative and theoretical. He was also unassuming, agreeable and intuitive. He felt deep compassion for his family and friends, and for the less fortunate people he met in his travels around the world. He was talented at handling people, too, as well as emotionally expressive; he shared his feelings. Darwin was also good with words; he wrote every morning of his adult life. But most telling, Darwin thought the way Negotiators think. He saw the big picture: he thought contextually, holistically, synthetically.

Web Thinking: The Searchlight

Psychologist Howard Gardner likened the synthesizing mind to a "searchlight" rather than a "laser" and saw this ability as an art. I call it *web thinking,* the natural ability to collect many bits of data as you think, weigh the importance of these variables and envision the relationships between them—the ability to think in webs of factors, not straight lines.

Negotiators find patterns in the data they collect. They read between the lines. They assemble facts, combine categories, unite opposites and focus on the larger concepts and principles these data suggest. They like tangents and intellectual diversions, too. In my Personality Type Study, only Negotiators overwhelmingly agreed with the statement "When reading, I enjoy it when the writer takes a sidetrack to say something beautiful or meaningful."

Builders can collect facts for their own sake. But if a Negotiator has to memorize or work with unrelated facts, he or she gets bored and frustrated. Negotiators feel intellectually stifled when they can't reflect on the broader meanings of their data. They generalize and synthesize.

Estrogen: The Mother Lode

The Negotiator's web thinking stems primarily from the activities of estrogen—in both sexes. Like women, men inherit specific levels of estrogen; and some men express nearly as much estrogen as testosterone. As the writer Susan Sontag put it, "What is most beautiful in virile men is something feminine; what is most beautiful in feminine women is something masculine."

There is much chemical overlap between the sexes. In fact, in my Personality Type Study, 29.1 percent of individuals were Negotiators. But significantly more women than men were Negotiators, 35.8 percent and 20.4 percent, respectively. And as with my data on Directors, I was pleased to see this gender difference; it supports my proposition that the traits of the Negotiator are linked with estrogen. The hands of Negotiators, both women and men, also show signs of elevated estrogen activity. Negotiators are far more likely than other personality types to have second and fourth fingers of equal length or a longer second finger—signs that they were exposed to more estrogen in the womb.

But how does estrogen enable Negotiators to do their web thinking?

With brain wiring. The brain's two hemispheres are connected by millions of nerve fibers, and individuals who were exposed to more estrogen in the womb have more connections between these brain halves. Estrogen also builds more nerve connections between distant regions within each hemisphere. In short, fetal estrogen constructs a well-connected brain—contributing to the Negotiator's ability to collect and integrate a broad range of information.[1]

Puberty and life experiences also adjust estrogen activities, however, altering web thinking—as two sex-change patients discovered. After a few months of estrogen injections, a man who was turning into a woman reported, "Whereas once as a man it had taken me seconds to make a decision, I would think things through much more carefully, weighing up all the options before deciding what to do." Moreover, after three months of testosterone injections to transform into a man, another individual said, "I do miss the overall picture." In this case, testosterone had suppressed estrogen activity and stolen the Negotiator's view.

Which brings me to an important point. Estrogen and testosterone can suppress or enhance each other, depending on many circumstances and interactions.[2] But you inherit your basic levels of each. Football players, for example, often have high levels of both testosterone and estrogen. Former U.S. president Bill Clinton must have a good deal of estrogen, too, because he thinks holistically. Indeed, he values web thinking. As he once said, "Presidents need to have what I could call a synthesizing intelligence."

This synthesizing intelligence has many perks—among them is imagination.

Imagination

I have come to believe imagination is a slightly different brain activity than creativity. People with powerful imaginations *combine* existing data and ideas, while those who are especially creative *generate* new data or ideas. We often do both, of course. But I suspect that imagination stems from web thinking. For what is imagination but the ability to reach into the breadth of one's stored knowledge, mentally assemble reams of information in new ways and "suppose" how these combinations of variables connect and interact?

Soul; self; love; truth; potential: Negotiators like to think abstractly, to think about things that can only be imagined. So to measure imagination, I included three statements in my Personality Type Study: "I vividly imagine both wonderful and horrible things happening to me"; "I frequently catch myself daydreaming"; and "When I wake up from a vivid dream, it takes me a few seconds to return to reality." Negotiators were the only type to heartily agree with all three statements.

Darwin imagined constantly. While in Chile, he experienced a violent earthquake, caused by the eruption of three volcanoes in the Andes. Darwin wrote, "The earth, the very emblem of solidity, has moved beneath our feet like a thin skin floating on liquid." He vividly imagined what geologists would discover decades later: plate tectonics. But his most imaginative moment came as he pondered the finches on the Galápagos Islands. The finches on each island were noticeably different from one another. It was this observation that stimulated him to propose his theory of natural selection by survival of the fittest. As he wrote in 1839, "Here, both in space and time, we seem to be brought somewhat near to that great fact—that mystery of mysteries—the first appearance of new beings on this earth."

Imagination is associated with estrogen. The sex-change patient mentioned earlier also reported that after three months of testosterone injections, she "imagined less," her "fantasy life had diminished strongly," and her thinking was becoming "more concrete." Indeed, scientists are beginning to understand how estrogen affects imagination—through the memory system. The brain's primary memory center, the hippocampus, binds facts and other data to build coherent scenes of both the past and future. And the hippocampus is highly sensitive to estrogen.

"Genius, in truth, means little more than the faculty of perceiving in an unhabitual way," wrote psychologist William James. At thinking in unhabitual—imaginative—ways, Negotiators shine. They're built for it.

Intuition: The Art of *Chunking*

Negotiators have another intriguing quality: intuition.

Sherlock Holmes, the astute detective in Sir Arthur Conan Doyle's crime novels, described this thinking process in himself, saying, "From long habit the train of thoughts ran so swiftly through my mind that I arrived at the conclusion without being conscious of the intermediate steps." Doyle had perfectly described the basis of intuition: *chunking*.

As a person learns a body of information, he or she begins to recognize its patterns and organize them into blocks of knowledge, the essence of *chunking*. With time, more patterns are chunked and linked, and these clusters of knowledge are stored in long-term memory. Then, when an experienced person sees a single detail of a complex pattern, he or she rapidly recognizes the larger picture, skipping all the steps of plodding sequential thought.

For example, if you know someone well, you are likely to know what that person is thinking from his or her wink or half smile. Intuition is the ability to reach into one's bin of stored knowledge and quickly and accurately perceive a complex pattern from a tiny bit of data. Like imagination, intuition is a spinoff of the contextual, synthesizing, web-thinking Negotiator's brain.

Gut Feelings: Body Loops

But when Negotiators intuit, they also *feel* that they understand. They'll say, "I know it in my gut."

Josh, a former world-class chess champion, described this sensation, writing, "Often in chess, you feel something is there before you find it. The skin suddenly perks up, senses heighten like an animal feeling danger or prey. The unconscious alerts the conscious player that there is something to be found, and then the search begins. I start calculating, putting things together."

"Gut feelings" are real feelings, the handmaidens of intuition. The brain is well connected to the skin, stomach, heart, lungs and other bodily organs via *body loops,* neural circuits that register "gut reactions" as someone gets a "hunch." Estrogen facilitates these body loops.[3] As a result, Negotiators may feel more bodily sensations as they think—biological cues that help them to intuit.

Negotiators have confidence in these insights, too. As Oprah Winfrey put it, "I am almost completely intuitive. The only time I've made a bad business decision is when I didn't follow my instinct." If Explorers trust their impulses, and Builders trust their values, and Directors trust their logic, Negotiators trust their intuition.

Mental Flexibility: Tolerating Ambiguity

"The test of a first-rate intelligence is the ability to hold two opposed ideas in the mind at the same time, and still retain the ability to function," wrote F. Scott Fitzgerald. Negotiators also excel at tolerating ambiguity. And estrogen is probably involved—because studies have shown that women, on average, are especially adept at holding two or more *opposing* ideas in the mind at once. Still, male Negotiators also have this ability.

Darwin, for example, appears to have been comfortable with ambiguous feelings about the existence of God. In his famous last paragraph of *The Origin of Species* he wrote, "There is grandeur in this view of life [evolution], with its several powers, having been originally breathed by the Creator." Historians quibble over Darwin's beliefs about religion, and his motives for referring here to the Creator. Darwin regarded himself as an agnostic. Yet he did more to undermine belief in God than any other person on this planet. Indeed, he once mockingly called himself the "devil's chaplain." Darwin may have held within his mind two dramatically opposing concepts: those of evolution and a possible Creator.

Imagination, intuition and the ability to tolerate ambiguity all stem from web thinking. Most likely so does the Negotiator's mental flexibility. In my Personality Type Study, Negotiators were far more likely than other types to agree with the statement "I can change my mind easily." In fact, this mental flexibility is directly linked with a gene or gene cluster on the X chromosome and associated with the activity of estrogen.

People People

Another outstanding trait of Negotiators is their curiosity about people. In my Personality Type Study, only Negotiators heartily agreed with the statement "I like to get to know my friends' deepest needs and feelings." They also strongly endorsed the statement "I highly value deep emotional intimacy in my relationships."

To Negotiators, everyone is unique, special and important. They thrive when they can build intimate attachments and share their lives with others. In fact, Negotiators want to create some form of intimacy with just about everyone they meet—whether for only a few moments of light conversation in the supermarket or a lifetime together as colleagues, friends or partners. When a relationship is not personal, authentic and caring, Negotiators are not interested in it.

I recently experienced this response myself when I spent the evening with a girlfriend. Before visiting me she intimated on the phone that her husband had been extremely sick, that she was facing financial ruin and that she would tell me more when she arrived. But as she settled on my couch, she announced that she didn't want to talk about "it." I objected; since she had already piqued my curiosity and my concern, I felt I had the right to know her secret. But she was adamant. So we spent the next four hours engaging in "small talk." She departed still unwilling to discuss her situation.

Reflecting on this encounter, I think I know what happened. Because my friend is a BUILDER/Director (according to her scores on my personality test), she probably found our chitchat relaxing, even intimate; and as a Director, she most likely expected me to respect her right to privacy. But to me, an EXPLORER/Negotiator, the evening was a waste of time. As an Explorer, I wanted to have my curiosity assuaged; as a Negotiator, I felt expelled from her personal world.

Things Left Unsaid: Connecting

Negotiators must connect on a deeply personal level, as I could see among Negotiators on Chemistry.com. On this site, I featured the cover of a book showing a man and woman on a sunny balcony overlooking a distant field. The man in the photo is walking away, and the woman is watching him go. Both have their back to the viewer. I asked members of Chemistry.com, "If you were the publisher and had to choose a title for this book, what would it be?" They were given four selections. Explorers picked the title *Adventures on the Rhine;* Builders chose *Anatomy of Friendship;* and Directors preferred *Power Plays.* But Negotiators chose *Things Left Unsaid.*

This drive to connect is linked with estrogen and oxytocin—a chemical produced, stored and triggered largely by estrogen. Women tend to have more oxytocin activity than men. And a great deal of data show that, on average, most women are more interested in cooperation and interpersonal harmony than most men. They more regularly cast themselves in a web of friendships, then they work to keep these ties intact.[4]

Many men express these traits, too, of course. Darwin was deeply connected to his family and friends. He saw them regularly. But he also felt a close connection to all living creatures. As he once wrote, "If we let imagination run wild, then animals, our companions in work and amusement, and ourselves may be one—we may all be netted together."

Eager for these connections, Negotiators strive to smooth out differences in rank. Negotiators seek consensus. They want to agree, to accommodate, to live in supportive networks unfettered by competition and discord. Controversy, disputes and debates make them uncomfortable. In my Personality Type Study, Negotiators did not agree with the statement "Debating is a good way to match my wits with others." They were also the least likely to concur with the statements "I am tough-minded" and "I think it is important to be direct."

Negotiators want to please. When these men and women make choices, they consider the needs of everyone involved. Once again, girls are a good example. When girls play informally, they rarely engage in contests with clear winners and losers. They form *flat packs,* small, nonhierarchical, leaderless groups. If a conflict erupts, they change the rules, make exceptions or stop the game; most girls want to preserve social harmony. And as estrogen levels increase with puberty, the urge to connect and cooperate intensifies.

Negotiators live in a world of win-win, not win or lose. They want to be liked. And they have an arsenal of skills to ensure that they can connect with others and act appropriately when they do.

Empathy

Foremost among these gifts is empathy, a hallmark of the Negotiator. In my Personality Type Study, Negotiators led the way in their positive responses to the statements "I am very empathetic" and "I am very sensitive to people's feelings and needs." Even their doodles show this compassion. Negotiators often draw hearts.

Empathy has two basic aspects. First is the ability to accurately infer what someone else is thinking and feeling, an aptitude known as *theory of mind.* Second is the drive to respond appropriately to another's needs and thus connect with that person. Both traits have been associated with estrogen and oxytocin.[5] In fact, as oxytocin activity rises during stress, it tends to trigger a response in women known as *tend and befriend;* they begin to comfort and nurture others.

Darwin was enormously empathetic. On board the HMS *Beagle,* he got into terrible arguments with Captain Robert FitzRoy about slavery. FitzRoy believed these Africans preferred to live under the protection of a master on a plantation. Darwin resolutely maintained that no human being would choose to be a slave; he clearly felt genuine compassion for slaves. He also demonstrated compassion as a young man when he dropped out of medical school after observing an operation being performed on a young boy. Darwin couldn't tolerate the child's excruciating screams. He fled the room and never returned to pursue medicine.

No Man Is an Island

"I feel your pain." It's an old line. But some people—particularly Negotiators, I believe—actually do feel your pain, due to *mirror neurons.*

These cells "fire" (or shoot off signals) in the observer when he or she watches someone else do or feel something. For example, pain cells in the brain normally fire in someone when you poke them with a needle. But when some people watch another person being poked, pain cells in *their own brain* also fire. More important to this book, individuals who score higher on tests of empathy also show higher activity in these mirror neurons; hence some scientists call these the *Gandhi neurons,* after the highly compassionate Indian leader. I suspect Negotiators have a particularly sensitive "mirror neuron" system.

Mimicry encourages empathy, too. When you see someone smile, your face naturally mimics that facial expression, if only fleetingly. And as you pull your muscles into this friendly grin, you trigger nerves that jump-start the *same* brain chemicals that are active in your smiling comrade, enabling you to feel some of the pleasure that he or she feels. Negotiators are natural copycats, and their mimicry most likely intensifies their empathy.

Imagination contributes to empathy, too. When people watch a video of patients in pain after a medical procedure, those who imagine themselves in the same situation show more activity in their own pain centers.

"No man is an island unto himself," wrote poet John Donne. We are built to reach out to others and feel compassion. Empathy is social glue. But because it is linked to estrogen and oxytocin, empathy is probably most common and intense in Negotiators.

Executive Social Skills

Negotiators have an arsenal of other social talents, all due to the activity of estrogen and oxytocin.

For example, these chemicals give you the capacity to override the impulses that distract you from completing your social goals, as well as the ability to read another's postures and gestures and to recognize someone's emotions from looking at his or her face.[6] "A face is . . . a whole cluster of faces, juxtaposed on different planes so that one does not see them all at once," mused Marcel Proust. Indeed, we make many complex facial expressions to convey some 412 different emotions. But those who primarily express estrogen, both women and men, excel at picking up subtle facial cues.

Lilt, inflection, cadence, pitch—Negotiators are also attuned to the music of your voice.

An acquaintance of mine with particularly acute social sensitivity is Josh, the world-class chess player I referred to earlier. Because he is so good at chess, a highly spatial game, Josh is most likely predominantly a Director. I suspect that his secondary type is the Negotiator, however, because of a story he told me.

He was about to play a major chess match in the Caribbean, and during the lunch hour prior to the game, he was standing on the terrace of the resort hotel, watching his opponent stroll in a nearby garden. A storm was brewing. Soon the sky erupted, pelting warm water on all below. Josh studied his adversary. If his opponent dashed for cover, Josh reasoned, he might be the kind of fellow who didn't like the unexpected; if his opponent sauntered back to the terrace delighting in the deluge, he might crave novelty instead. Josh told me he would play the game accordingly. He'd keep his opponent off guard with wacky moves if his rival was flustered by the unusual. He'd play a highly conventional game if his foe yearned for excitement.

Josh displayed astute *theory of mind* and executive social skills, traits of the Negotiator. He also won the chess match.

Agreeableness

Negotiators are agreeable. Agreeableness is a constellation of many traits. Scientists report that agreeable people are found around the world, describing them as sympathetic, cooperative, compliant, considerate, charitable, forgiving, altruistic, trusting and warm rather than aggressive and cold.

These men and women also tend to be authentic, friendly, diplomatic, accommodating and polite. In fact, in my Personality Type Study, Negotiators were the least likely to agree with the statement "I am tough-minded." They also disagreed with the statement "I like to avoid the nuances and say exactly what I mean."

A highly agreeable Negotiator is Lisa, a professional matchmaker. I had always wanted to meet a matchmaker, so I welcomed the opportunity to meet her recently in a coffee shop in New York. I knew she was a Negotiator the moment I saw her. She walked in with an inquisitive demeanor, radiating curiosity about me. She looked me straight in the eye with an attentive "anchoring gaze," listened carefully and responded with warmth and sincerity. She even used the words typically used by the agreeable Negotiator, including *relate, connect* and *feel*.

Estrogen must be the sap of the agreeable disposition, because transsexuals who inject estrogen to transition from male to female become less confrontational. Instead, they employ indirect aggression, such as sarcasm, irony and gossip. Moreover, girls and young women are, on average, far more agreeable than boys and men. More girls apologize for no apparent reason. Girls and women give more compliments. Girls interrupt less, tease less, brag less, threaten less and criticize more softly. And girls and women rarely blurt out, "You're wrong."

Darwin was particularly courteous and self-effacing; he hated to offend. In fact, his agreeableness nearly cost him his lifework. Darwin knew that his evolutionary theories would topple many of the basic tenets of paleontology, anatomy, botany, geology and other sciences, as well as seriously challenge current beliefs about God. So he worked for twenty years in secret, telling only family and a few close friends about his "big book" on the origin of species. He didn't want to upset his religious friends and relatives, particularly his beloved wife, Emma Wedgwood.

Only when Alfred Russel Wallace, another brilliant naturalist, independently arrived at the same idea of natural selection by survival of the fittest did Darwin act. Wallace had come to this idea during a malarial fit in the Moluccan Islands. He wrote a short paper outlining his novel hypothesis and sent it to Darwin. Upon opening this mail, Darwin panicked, writing to a close friend, "All my originality . . . is smashed." Then, despite his years of painstaking research to support this idea, Darwin included Wallace as codiscoverer on their 1858 joint paper—the first presentation of an idea that would rock the world.

Trusting

To call someone trusting is not always a compliment, since some regard those who easily trust as weak and vulnerable. Yet a trusting nature can be an asset. If you trust the right person at the right moment, you can save yourself time and energy, as well as build crucial social connections. Trusting others is an adaptive social skill—and yet another hallmark of the Negotiator. If Explorers are optimistic, Builders pessimistic, and Directors skeptical, Negotiators are trusting.

Trusting has been associated with oxytocin, the chemical triggered by estrogen. In fact, in one experiment, men who had been given oxytocin to sniff just before playing a game for money began to express more trust in the stranger they were playing with; they also began to cooperate more with this unknown individual to win.

Perhaps this link between trust and oxytocin explains why some people take the drug Ecstasy. Ecstasy elevates levels of oxytocin, serotonin and dopamine. It is the oxytocin, scientists believe, that triggers this "prosocial" experience—the feelings of empathy and connectiveness, as well as the drive to trust.

Emotional Expressivity

"It is as healthy to enjoy sentiment as to enjoy jam," remarked British writer G. K. Chesterton. Negotiators like to express their feelings. These men and women are often emphatic about their feelings, too. *Totally* enraptured; *completely* engrossed; *perfectly* thrilled; *absolutely* fascinated: their words are often emphatic and dramatic. We all feel an incredible range of emotions, often with piercing intensity and persistent regularity. But in my Personality Type Study, Negotiators were far more likely than other types to agree with the statement "I feel emotions more deeply than most people."

All is not happiness for Negotiators, however. Sorrow, self-doubt and many other feelings often haunt these men and women. Bill Clinton recalled the summer afternoon in 1963 when, at age seventeen, he settled in the big reclining chair in his family's den to watch Martin Luther King Jr. on television. King was standing in front of the Lincoln Memorial in Washington, D.C. And in the rhythmic tones of a Negro spiritual, he delivered his "I have a dream" speech, calling for freedom and brotherhood in America. "I started crying during the speech and wept for a good while after Dr. King finished," Clinton wrote.

For Negotiators, emotions linger. When I asked each type to

respond to the statement "After watching an emotional film, I often still feel moved by it several hours later," Negotiators were far more likely to agree. In fact, women have a better memory for emotional experiences, and women express more estrogen.

Teenage girls provide a stunning example of the link between estrogen and emotional expression. As estrogen floods the teenage brain, some young women turn into "drama queens"; they overdramatize everything they feel. Men can be drama queens, too, of course. I know several male Negotiators who react to minor incidents with tremendous joy or deep hurt. Every personality type expresses strong emotions upon occasion, but Negotiators leak their emotions all the time.

And Negotiators are particularly good at expressing their emotions with words.

The Gift of Gab

Verbal fluency has long been associated with estrogen. In fact, verbal fluency—the ability to find the right words rapidly—as well as the capacity to pronounce words and remember words increases during midcycle ovulation, when estrogen levels peak. But many men—primarily Negotiators—also have remarkable verbal aptitudes. In my Personality Type Study, for example, both male and female Negotiators were the most inclined to read books. In fact, among the top ten words Negotiators use when they describe themselves are *read* and *reader*. And when Negotiators talk and write, they can be particularly poetic, using colorful imagery and gripping prose.

For example, Darwin's description of the "tangled bank" in his final paragraph of *The Origin of Species* is much loved among scientists. Speaking of all living creatures, he wrote, "It is interesting to contemplate a tangled bank, clothed with many plants of many kinds, with birds singing on the bushes, with various insects flitting about, and with worms crawling through the damp earth, and to reflect that these elaborately constructed forms, so different from each other, and dependent upon each other in so complex a manner, have all been produced by laws acting around us . . . and that, whilst this planet has gone cycling on according to the fixed law of gravity, from so simple a beginning endless forms most beautiful and most wonderful have been and are being evolved."

Superconsciousness and Altruism

Of all the traits of Negotiators, perhaps the most remarkable is their exquisite awareness of everything around them.

Like Darwin, Russia's great novelist Leo Tolstoy displayed this awareness vividly. As biographer A. N. Wilson wrote of him, "One of the things that makes him such a memorable writer is his extra-consciousness, or super-consciousness, of existence itself . . . this pre-ternatural ability to be aware . . . This super-awareness came to him, he informs us, when he was still a baby."

With their web thinking, imagination, intuition, flexibility, social skills and empathy, Negotiators are often superaware of the entire living world.

Moreover, they are driven to give back, to contribute to society at large. If Explorers strive to make an impact, Builders crave to belong, and Directors need to achieve, Negotiators must help. So they adopt causes and choose careers to improve human welfare. In my Personality Type Study, for example, nurses were most likely to be Negotiators. I also found that Negotiators fraternize with people like themselves, other activists, those who share their superconsciousness and are striving to help humankind in one way or another.

But the Negotiator's form of altruism is different from that of the Director. Whereas Directors are inclined to dash into a burning building to help a stranger, a trait linked with elevated activity of testosterone, Negotiators clean up rivers, raise money to fight cancer or donate their technical skills to help the poorest of the poor. Negotiators think ahead, weigh the odds, plan their strategy and rarely put themselves directly in harm's way—a type of altruism that has been associated with *low* levels of testosterone. Estrogen and oxytocin are probably involved in this trait, altruism, since these hormones are linked with nurturing in many species.

An outstanding example of an altruistic Negotiator is a friend of mine, Paul, an Irishman. Paul was raised a Christian but became a Muslim in his young adulthood. Paul has devoted his life to helping others, particularly New York Muslims who are victims of discrimination. But even as a child Paul had an extraordinary social awareness. Soon after he read Roald Dahl's story "Pig," he gave up eating meat. He hasn't worn a leather belt, leather shoes or a leather coat or used a leather suitcase since he was twelve. Like Darwin, Tolstoy and probably many other Negotiators, Paul is "superconscious" of every sentient being on this planet. This consciousness fuels his altruistic nature.

"The value of a sentiment is the amount of sacrifice you are prepared to make for it," wrote the English novelist John Galsworthy. Many Negotiators sacrifice just about every day. Only Negotiators in my Personality Type Study strongly agreed with the statement "Regardless of what is logical, I generally listen to my heart when making important decisions." Negotiators do what is most important for the group, even if it is impractical, inconvenient or expensive. In fact, of the four types, Negotiators, both men and women, are the most likely to choose low-paying jobs, such as social work.

Charity is a supreme virtue to Negotiators. They help because they must. Indeed, I suspect their self-esteem is greatest when they are helping others. And how Negotiators feel about themselves is crucial to their well-being.

Introspection

"Applicants for wisdom do what I have done: inquire within." So spoke the fifth-century Greek philosopher Heraclitus. Not only do Negotiators look out; they also look in. Self-knowledge is essential to the Negotiator. They want to travel into the deepest byways of their psyche to find their essence and learn who they truly are. For them, self-discovery is a quest. As Oprah Winfrey succinctly put it, "There is no moving up and out into the world unless you are fully acquainted with who you are." Negotiators also believe they must *become* this essence. They want to be authentic instead of presenting a facade.

But here's the rub: because Negotiators have such a nuanced awareness of their every thought and action, they often live with the uncomfortable sense that they are not being genuine. I understand this. As a Negotiator, I sometimes mutter to myself, "Will the real Helen Fisher please stand up?" Alas, a committee would rise. Negotiators often feel they are more than one person. So they search for their center and their place in nature. Discovering their identity is paramount to the Negotiator.

On the Dark Side:
Negative Traits of the Negotiator

Each of these four personality types can be unpleasant, of course, even unbearable at times. So it is with the Negotiator.

Their broad, contextual web thinking can make Negotiators *system blind*, unable to focus on the important details while they ruminate on the bigger picture. Some digress so often that they appear scatterbrained. Because they can see so many angles to any issue or decision, they can be indecisive, too, mulling over all the possibilities instead of acting. Because Negotiators camp in the forest of big ideas, they can also be absentminded about everyday chores.

Their emotional expressivity can make them appear effusive. Their need to please others can turn to placating. And they can be overly trusting, becoming gullible. Because they refuse to confront, some Negotiators will stab you in the back. And they can be particularly vicious with words.

Negotiators are skilled at talking, listening and connecting emotionally with others. But many are so socially savvy that they can be glib. Some can be tediously verbose. And some use half sentences or rapidly change the topic, making the conversation disjointed. Indeed, I have seen two Negotiators maintain a steady dialogue during which neither ever stopped talking. Both were continually talking and listening at the same time.

Negotiators can appear nosy because they are so interested in others. They can get overly involved in your personal problems, too, then complain because they feel burdened by their role as sounding board and mentor. Because Negotiators seek connection, some appear needy, constantly expecting you to reassure them that you are interested and concerned. They can also project their own feelings and attitudes onto you, assuming that you see the world their way. Negotiators can also take casual comments as criticisms and dwell on these perceived slights. And when they feel betrayed, they can be unforgiving, holding a grudge for months or years.

Negotiators want to help, a noble trait. But they can carry their altruism too far, then feel hurt when others don't appreciate their self-imposed ordeals, and even become critical of friends or relatives who don't share their lofty goals. And as these men and women seek to help others, they also search to understand themselves. Some overdo it, though; they become self-absorbed, self-conscious and self-critical.

Finally, Negotiators may be particularly susceptible to depression. Surveys in over a dozen countries, including Canada, France, Germany, Italy, South Korea, Lebanon, New Zealand and the United States, report that women express more symptoms of classically defined clinical depression than men. I suspect these experts are really referring to both women *and men* who primarily express estrogen: Negotiators.

Philosopher Kings

Of all the Negotiators in recent history, among the most distinguished for his contribution to humanity is Mohandas Karamchand Gandhi—a supreme Negotiator.

Like all Negotiators, Gandhi saw the big picture. He was imaginative and intuitive. He spoke eloquently. He was enormously skilled at handling people. And he had a deep compassion for all humankind. He was also nonviolent, a core quality of the Negotiator. And with satyagraha, the Gandhi form of nonviolent resistance, he overcame a monumental force: the British Empire. Without bullets, tanks, generals or even soldiers, Gandhi almost single-handedly brought political freedom to some 500 million men and women of India in the mid-twentieth century. All he had was a philosophy.

Plato called this type of human being the Philosopher King—the peaceful, virtuous, temperate, charitable, authentic man or woman who strives to understand the meaning of life and create a better world. What do Negotiators achieve with their broad gaze outward and their penetrating search within?

Wisdom.

To thine own self be true,
And it must follow, as the night the day,
Thou canst not then be false to any man.

—WILLIAM SHAKESPEARE, *HAMLET*

PLAY MATES, HELP MATES, MIND MATES, SOUL MATES:

The Chemistry of Dating

Many years ago I was walking toward the Lexington Avenue subway in New York when I found a quotation scrawled in blue chalk on the pavement. It read, "We struggle all our lives to know a few basic things about ourselves that the most casual passerby could tell us in a minute." Ever since, I have wondered what my friends, relatives and acquaintances could tell me about myself. Now, as I write this book, I am acquiring what I call a PGO—a penetrating glimpse of the obvious. I hope you are learning more about yourself as well.

But how does our basic biological nature color our search for "him" or "her"? How does each type woo? And what kind of love is each type looking for in a mate?

The ancient Greeks had more than ten words to distinguish different types of love. Sociologist John Alan Lee reduced these overlapping categories to six. Most celebrated is *eros*—passionate, sexual, erotic, joyous, high-energy love for a very special person. *Mania* is obsessive, jealous, irrational, possessive, dependent love. *Ludus* (rhymes with *Brutus*) derives from the Latin word for play; this love is playful, unserious and uncommitted. *Storge* (rhymes with *more gay*) is an affectionate, brotherly or sisterly kind of love, a deep feeling of special friendship that lacks a display of emotion. Gentle, unselfish, dutiful, all-giving, often spiritual love is known as *Agape*. Last is *Pragma,* love based on compatibility and common sense, pragmatic love.

Everyone who falls intensely in love feels some degree of Eros and Mania. Romantic love is an irrational, passionate, erotic, possessive, exhilarating obsession. But the Explorer, Builder, Director and Negotiator tend to search for more than this; each also seeks a different variation of these Greek forms of romance as well.

For example, until they are ready to build a serious long-term relationship, Explorers tend to express more *Ludus*—playful, casual, uncommitted love.

The Explorer: Looking for a Play Mate

Explorers seek someone who will go adventuring with them, in conversation, in bed, around town or around the world—a "play mate."[1]

Explorers regard courting as entertainment. They watch less television than other personality types. Instead, they prefer to go out and do novel things. And they have many interests. So dating an Explorer can be exciting. A striking example is an Explorer friend of mine who visits me occasionally when he comes through New York from Europe.

Recently he came for the evening, so I assembled a number of things we could do. Three possibilities seemed especially appealing to me: a lecture on "Einstein: His Life and Universe" at the New-York Historical Society at 6:30 PM; a one-man recital of Tennyson's long poem "Maud," performed in a tiny theater in Greenwich Village at 8:00 PM; or dinner at the steak house Keens, where the ceilings are covered with antique Dutch clay pipes and other memorabilia from old New York.

I asked him which of the three he preferred.

"All of them," he replied.

We did all three. Then he took me home in a pedicab, whizzing through the glittering city at midnight in a tiny, open carriage drawn by a man madly pedaling his bicycle.

"Life is like riding a bicycle," wrote Albert Einstein. "To keep your balance you need to keep moving." Explorers keep moving, physically, intellectually or both. They often have boundless energy, a trait associated with dopamine and norepinephrine—the predominant neurotransmitters of the Explorer.

Explorers are highly sensitive to the world around them, particularly colors, lines, forms and textures. Indeed, their attraction to complex designs and abstract art has been associated with increased dopamine activity and was common among the Explorers at Chemistry.com.

Explorers like the pleasures of the senses. So they are inclined to take a date to all sorts of arts and cultural events—from museums and art galleries to food festivals, parades, wine tastings, sound and light shows, scary movies, poetry readings, design or fashion shows, aroma bazaars or chocolate factories. Explorers are also more likely than other personality types to enjoy spicy, sour and crunchy foods, as well as foreign foods. As a result, many are just as happy taking you to an Ethiopian or Turkish bistro as to a fashionable restaurant.

Explorers also gravitate to music that is intense and complicated, preferring classical, jazz and hard rock to movie sound tracks and other simpler tunes. These men and women move with a sense of rhythm, too; most are comfortable in their skin. Even the Explorer's language is more rhythmic and lyrical than that of other types; it's vivid, original, vivacious and often full of slang.

Most Explorers have had a good deal of experience at dating. They have a tendency to play the field and seek as much freedom and variety as possible. So Explorers can adapt to just about any courtship situation; they blend in. And because they are friendly and enthusiastic, and have little interest in rules or schedules and no desire to control others, they can make a date feel comfortable quickly. Explorers are also sponta-neously generous, often giving presents early in a relationship.

Explorers are good at listening and talking. They are inquisitive, and likely to ask you about yourself—to engage you as well as satisfy their curiosity. Many Explorers gaze directly at you when they con-verse, leaning forward, pausing less, responding quickly when spoken to and smiling and laughing readily. And most Explorers move easily from one topic to another in an open-ended, wide-ranging, freewheel-ing exchange.

Explorers find self-disclosure difficult, however. They are likely to avoid conversations about their feelings, deflecting them with humor.

Nonsense and incongruous humor and amusing situations that remain unresolved are all particularly appealing to Explorers. Perhaps because the action is sustained. Explorers easily laugh at themselves, too. They are often good at one-liners and quick, witty exchanges. Many engage in occasional antics, such as mimicking others or leap-ing or dancing in the street. And, like Negotiators, they are people pleasers. So Explorers entertain, even perform. Time with an Ex-plorer can also be laced with surprises, irony and pranks, even silli-ness. I have one Explorer friend, for example, who spontaneously leaps over garbage cans in the street, even though he looks like an ur-bane white-haired gentleman.

"The only unnatural sexual behavior is none at all," Sigmund Freud declared. Explorers agree; they are the most sexual of the four types. Explorers have a relaxed attitude about sex, although Explorer women are less permissive than men. In fact, Explorers often regard sex as recreation, a natural function that is healthy but not necessarily linked with loving—or even really liking—their partner.

Add their taste for novelty, their impulsivity, their charm and their lack of self-consciousness, and (as you might expect) psycholo-gists have found that Explorers have more sex with more partners than other types. Moreover, after marriage, female Explorers continue to have more marital intercourse. Interestingly, my Personality Type Study showed that Explorers (and Directors) are also more willing to

discuss their sexual needs with a partner than are Builders and Negotiators. And Explorers are the most likely to agree with the statement
"Sex is an essential part of a successful relationship."

Are Explorers thrill seekers in the bedroom?

Probably. In fact, variations in the dopamine system are directly
associated with a higher sex drive and more varied sexual experiences.[2]

But Explorers vary. One's secondary type plays a powerful role in
who you are and how you love, as do many other biological and cultural forces. For example, for most Explorers, morality is a flexible concept; ethical stances must reflect the needs of the specific situation. But
EXPLORER/Builders are likely to have a more fixed moral compass
and seek a partner who shares their view of duty and tradition. This
type is also more likely to want to "do the right thing." So they follow
more conventional rules of courtship, like bringing flowers and being
on time. These men and women like closure, too; they want to know
where they stand in a relationship. EXPLORER/Builders gravitate to
mates who are traditional and orderly—except when an adventure
comes along.

Dan, the president of a major American firm that sells condi-
ments and foods, is a good example of an EXPLORER/Builder. Dan began working for his father after class while still in high school,
combining sugar and cola syrup and selling this mixture to local soda
fountains in Brooklyn, New York. From this beginning, he built a
world-class company, using his creativity and daring. Dan is a risk taker,
an Explorer. He is also a voracious reader of history, literature and
world affairs, and he collects ancient Greek art—more Explorer traits.

But Dan is politically conservative, as well as a hands-on
manager—Builder traits. He is also modest, calm, affable and
social—more Builder traits. Perhaps most telling, he married a dynamic
woman, as Explorers do. But as he said bashfully of the moment he
became interested in her, "Well, I was going out casually with several
women at the time. But what really impressed me about Leslie was
when she came over with her homemade cookies. It was all easy sail-
ing after that." Spoken like a traditional Builder.

In short, Dan is a high-functioning mix of both dopamine and serotonin—an EXPLORER/Builder.

EXPLORER/Directors are different. These men and women have nerves of steel. Like many Directors, they thrive on the edge; and they act fast—because they don't feel it's necessary to justify their actions. Moreover, nothing is sacred to the EXPLORER/Director, including courtship traditions. These men and women are probably most likely to elope. They must have a partner who can appreciate their appetite for autonomy and rank, however. These men and women are also the most sexual of the Explorers, because both dopamine and testosterone elevate sexual desire.

EXPLORER/Negotiators are the most diplomatic of the Explorers. They seek harmony. So on a date they tend to be accepting, flexible and sensitive to the feelings of their partner. They have the Negotiator's big heart. And because they are personable and easygoing, these men and women often have a wide range of suitors. Nevertheless, they are particularly attracted to individuals who share their live-and-let-live attitude. And they tend to avoid anyone who likes to confront, compete or overly schedule their time.

"Run my dear, / From anything / That may not strengthen / Your precious budding wings." In this fourteenth-century Persian poem, Hafiz captures the Explorer's view of life. Explorers find it difficult to commit—until they find a true play mate. Then an Explorer can be a thrilling, passionate and devoted partner.

The Builder: Looking for a Help Mate

"What a piece of work is a man!" exclaimed Hamlet in Shakespeare's classic drama, ". . . the paragon of animals!" In some ways we are the paragon of animals. One of humanity's supreme achievements, I believe, is our penchant to bear and rear our children as a team: pair-bonding, or monogamy. Only 3 percent of mammals form a pair-bond to rear their young. We are among them. And perhaps the most effective at *long-term* pair-bonding is the Builder.

Although just about everyone becomes passionate and ecstatic (and unrealistic) when they fall in love, Builders are the most likely of the four types to express what the ancient Greeks called *pragma*, love based on compatibility and common sense, pragmatic love. If Explorers seek entertainment, and Directors want a partner with a similar intellect, and Negotiators must have a deep personal connection, Builders seek a stable and predictable team player, someone who shares their fidelity to family and tradition—a "help mate."

For Builders, courtship is serious business. They believe in good manners, old-fashioned courtesy, punctuality and well-arranged schedules. Builders are likely to open doors for women, cook for men, be on time and make concrete plans for the evening. They send flowers, cards and candy. Builders are often highly protective of a date. And when they become serious about someone, they are likely to give important gifts, presents meant to have symbolic and enduring value. Builders particularly enjoy traditions, too, and can become sentimental easily.

Builders tend to think concretely; they are literal; they like and trust facts. So on a date, a Builder is likely to engage in detailed factual conversations—about food, clothing, prices, services, travel, sports, accidents, the weather, TV programs and/or movies. Builders are often good at small talk, short conversations on concrete topics. They can be very modest when discussing their achievements. And like Explorers and Directors, they are not generally introspective. So most do not appreciate a date who analyzes their feelings or behavior in exquisite detail.

Because Builders are uncomfortable with "drama" and other displays of emotional excess, they can appear unromantic to the expressive Negotiator, apathetic to the flamboyant Explorer or overly proper to the bold Director. But people often become attracted to a Builder when they are ready to settle down and begin a family. The Builder's clear-cut values, dependability and social grace become beacons.

Builders don't choose a partner impulsively. They are cautious.

And they tend to take the opinions of their friends and relatives seriously. Moreover, they are conscious of appearances, and they can be crushed by criticism. So they often want their friends and relatives to approve of their date before they begin to woo ardently.

As a result, Builders enjoy doing things as a couple within a larger group of friends. And Builders will bring a date to an "in" restaurant or some other place where they are likely to see their pals as well.

Builders want to belong, and to take in social, business and community events in style and comfort. They are likely to invite a date to a charity ball, dinner party or business reception. They especially like to attend sports events and to go to local attractions, such as zoos, amusement parks and historical sites and monuments. They enjoy these convivial outings, particularly with a group of friends, and feel they are also supporting the community. In fact, because many Builders are so involved in their social networks, they can fail to schedule enough private time with a potential mate. As a result, intimacy with a Builder tends to develop slowly.

This suits them. Builders like long engagements; this way they can meet their social obligations and think carefully about a possible marriage. Builders want romance to proceed as planned. They often have big plans, too. My Personality Type Study showed that far more Builders (and Negotiators) seek marriage or a committed relationship than do other types.

Despite peer pressure, Builders generally aren't interested in casual sex. For them, psychologists report, sex is not playful recreation, as it is for Explorers, but a way to forget their problems, get physical comfort and relax with a devoted partner. So Builders are the type most likely to "save" themselves for their wedding night. Some regard premarital sex as irresponsible, even morally wrong. As a result, Builders probably have fewer sexual experiences before marrying than do other types.

In fact, Builders are likely to be the least sexual of the four types. A great deal of scientific data has shown that elevated activity in the serotonin system can inhibit sexual desire. My Personality Type Study supports this finding. Builders are the least likely to agree with the statement "Sex is an essential part of a successful relationship." Builders are also less likely to be comfortable telling a partner about their sexual needs. Builders do, however, enjoy sex. They just take comfort in the familiar.

Like all types, of course, Builders are a blend of all four trait constellations, and their secondary temperament certainly plays a role in how they court and whom they choose to love. For example, although all Builders tend to be cautious, BUILDER/Explorers have an impetuous side; they seek novelty—within limits. This adds energy and freshness to their sensible temperament, catches their friends off guard and makes them both dependable *and* exciting on a date. In fact, they often fraternize with friends who are more adventurous, energetic and theoretical than they are. Yet like most Builders, these men and women are particularly drawn to someone who is trustworthy, caring and socially skilled—someone like themselves.

The BUILDER/Director has a streak of the independent, innovative thinker, comprising an unusual balance between conventionality and originality. Like Directors, BUILDER/Directors want a partner with whom they can have deep and focused conversations. Yet they are literal and emotionally contained. So these conversations must be concrete, factual and informed. BUILDER/Directors are probably the most ambitious and hardest working of the Builders, even on a date. So they are likely to take a partner somewhere that is difficult to find or get into. And they gravitate to partners with high moral standards, those whom they regard as honest and respectful of traditions.

BUILDER/Negotiators are the most socially graceful of the Builders. Confident and easygoing, they are likely to be the master of ceremonies at any social gathering. They particularly like to bring people together. So the BUILDER/Negotiator will quickly include a date in their social whirl and make a genuine effort to make these social occasions fun—carefully introducing him or her to everyone and telling friends about their date's interests and achievements. Because

BUILDER/Negotiators are relaxed and amusing, they attract a wide range of suitors.

"Strong is your hold O love," wrote poet Walt Whitman. Builders agree. They want a committed relationship, with a help mate. They seek someone who is sensible with money, responsible with possessions, respectful of traditions, industrious, interested in community and devoted to family life.

And when they find "the one," Builders nest—and take on all the responsibilities of a true partnership. They plan to be sexually faithful. And they strive to be the "good husband" or the "devoted wife." Indeed, Builders are willing to do time-consuming drudgery, going far beyond what is expected of them, to maintain their vows.

Usually, this isn't difficult. Psychologists report that these men and women are often highly satisfied with their partnerships.

The Director: Looking for a Mind Mate

For the Director, dating is not entertainment, as with the Explorer, or a social activity, as with the Builder, but an expenditure of time and energy. Directors approach dating the same way they approach their other interests, by rationally analyzing their hopes, needs and intentions. Foremost, they seek a partner who shares their goals, as well as someone who is eager to exchange ideas, build theories and talk about science, philosophy, politics, history or whatever interests them. They seek a "mind mate."

To balance out their forthright style and tendency to make decisions quickly, Directors also gravitate to people who weigh alternatives, listen actively, handle conflicts deftly and have other well-honed social skills. Because they analyze their own emotions, as well as hide them, Directors tend to seek mates who are emotionally expressive. Because they have difficulty dealing with "controlling" people, they tend to be attracted to those who are flexible, even indecisive. And they gravitate to individuals—often Negotiators—who have the empathy and intellectual complexity to handle their sometimes forceful personality.

Directors are rarely interested in pursuing several relationships at once; instead they want deep discussions on intellectual topics with a single partner. As a result, many have had little dating experience in their youth, and they can be socially awkward. For example, on a date a Director is capable of treating you with benign neglect if he or she doesn't envision a future with you. And if Directors find themselves in a dead-end relationship, they end it abruptly—before it gets out of their control.

Directors have a take-charge style. They like to debate, to spar. They can't tolerate moralizing. And they are impatient with anyone who doesn't interest them. But when a Director finds someone whom they regard as a person of quality and worthy of their time, they focus intently on him or her. They will talk and listen with rapt attention. Intellectual sharing is sexy, even romantic, to the Director.

And Directors can be very romantic. Sentimental crooning, sending routine cards or flowers, even anniversaries mean little to them. But they are prepared to make a deep and genuine commitment to another person. And when they find the right individual, they will carefully engineer heartfelt activities to do with their beloved. They do not like to engage in public displays of emotion, however, as they admire self-control. Instead, they focus on exactly what you want and try to provide it—even if it is a rare book or a trip to Pitcairn Island.

Directors also want to be needed, to help, to "fix" your problems. So if you clearly and logically describe your issues in the relationship or other difficulties, the Director will often go to the ends of the earth to help you resolve them.

In their romantic relationships, Directors seem to express more *Storge,* the Greek term for a deep feeling of special friendship that lacks the display of emotion. In fact, Directors generally have only a few serious relationships before they marry. In my Personality Type Study, I asked the question "How often do you fall in love?" Directors fall in love the least frequently of all the types. Yet once they have decided to build a partnership, they don't defect easily either, even if the relationship is not exactly what they had hoped for. Directors don't like to change their minds.

Directors have a high sex drive, a trait associated with testosterone. Most don't enjoy "sleeping around" while single. Once in a committed partnership, however, Directors regard sex as central to the union. When their sexual appetite is whetted, it can lead to more sex, too, because any kind of sexual activity triggers testosterone—creating a feedback loop of wanting sex, having sex, and wanting more sex. This doesn't apply only to men. High-testosterone women have more sexual fantasies than other women, as well as more sex.

Even when they are anticipating sex, testosterone levels increase. Known as the *lighthouse effect,* this response was first discovered when scientists measured the beard trimmings of a lighthouse attendant. He lived alone on an island during the week. But every Friday evening he sped to the mainland to enjoy the weekend with his girlfriend. On Fridays his beard grew more than usual—the effect of testosterone, triggered by sexual desire.

Directors seldom share the details of their sex lives with friends, however. If sex is entertainment for the Explorer, and relaxation for the Builder, it is deeply meaningful to the Director—woven with subtleties and symbolism. So the Director seeks a mate who is perceptive and imaginative and finds intimacy in this psychological journey. Directors, like Explorers, are also more comfortable talking about their sexual needs than other personality types.

Yet the Director's secondary type (as well as their childhood and other factors) will affect what they are looking for in a mate. Directors, as a rule, are highly independent; they follow only rules of their own making. And as they expect freedom in a relationship, they give freedom to their partner. But DIRECTOR/Explorers need the most privacy and autonomy. So they seek someone who is just as self-sufficient as they are, particularly someone who enjoys time alone. These men and women also like an irreverent mate, someone who shows little concern for social conventions, as well as a mate who is imaginative, adventurous and not competitive with them.

DIRECTOR/Builders are leaders, often world leaders. As Directors, they are highly motivated to succeed; as Builders, they are skilled managers with clear objectives. Even in courtship, these men and women know what they are looking for and rarely settle for less. They are also the most social of the Directors, gravitating to partners who will help them maintain their public, business and personal ties. DIRECTOR/Builders are the most likely of all Directors to enjoy the formalities of dating. They are pragmatic, skeptical and schedule-oriented, and they date purposefully. They can be hard to get to know, however, as they can hide their complex personality behind a wall of composure, logic and humor.

Directors follow few customs of courtship. But the DIRECTOR/Negotiator is especially unlikely to see any intrinsic value in courtship formalities. Even matrimony, although acceptable to DIRECTOR/Negotiators, is not as important to them as is their partner's personal conduct. Commitment and loyalty are paramount to these people, however. These men and women are also the most romantic of the Directors. They work to sustain this romance, too, even in a long-term partnership. They give special gifts, send special notes and celebrate special events with flair and sensitivity.

Directors are forthright. They let you know what they want and how they wish to proceed; and they can be hypersensitive to even tiny signs of rejection. With this personality type, you know where you stand. Moreover, Directors often know within a couple of meetings whether they can build a future with a date. And if they find "the one," most become dedicated mates—particularly when they regard that person as a "mind mate."

The Negotiator: Looking for a Soul Mate

Negotiators seek the ideal partner—someone with whom they can make an intensely intimate, deeply meaningful, inspiring and spiritual connection: a "soul mate." And when they find "him" or "her," they work tirelessly to promote harmony with this cherished companion, the love of their life.

Negotiators are popular with all types because they are flexible, appreciative, affectionate, good at reading others' thoughts and feelings, empathetic and superb at expressing their own ideas and emotions. So they tend to have a lot of dating choices. And because they are so curious about people, and tend to like and trust them, they are able to build a meaningful relationship quickly.

But Negotiators, like Directors, prefer to go out with one person at a time and to explore the depths of this potential partnership. If dating is entertainment to the Explorer, social participation to the Builder, and an investment of time and energy to the Director, it is the search for a deep and authentic personal connection to the Negotiator.

In fact, most Negotiators prefer to avoid social events when getting to know someone. Like Directors, they dislike wasting time on irrelevant, trivial or boring socializing; and as Negotiators, they can't tolerate chitchat in which no feelings are shared. In fact, to them parties are barriers to their quest to engage in heart-to-heart conversations—intimate exchanges that explore feelings, beliefs, goals and ethical dilemmas or the meanings behind art, music, poetry or other abstract topics. With these tête-à-têtes, Negotiators seek common intellectual and emotional ground and hope to build a bridge to another soul.

Because Negotiators reveal their hearts to a potential partner, they can be deeply hurt by rejection. In fact, Negotiators are often the type that stops dating for periods of time. To them, dating isn't "fun"; it is a true giving of themselves and a journey deep into the psyche of another—all for the high purpose of forming a bond for all eternity.

It's no surprise, then, that among the members of Chemistry .com, Negotiators, like Builders, are highly likely to say they are seeking marriage or a long-term committed relationship. This makes biological sense; both estrogen and oxytocin—the primary chemicals of the Negotiator—are associated with bonding, nurturing and attachment.

And when they find "the one," Negotiators are superb at generating and maintaining intimacy with this partner. Their emotional antennae are always quivering. They avoid conflicts with a mate and make major personal adjustments to strengthen this cherished bond. They give thoughtful gifts, such as a treasured book or photograph. They bestow meaning on everyday interactions, such as a ritual morning kiss or afternoon phone call at the office. They share their most personal feelings. And they want their partner to share his or her feelings too. In fact, Negotiators particularly enjoy analyzing their partner's thoughts and motives—as well as the things that stabilize or undermine the union.

Negotiators are the most romantic of the four types. In my Personality Type Study, I asked the question "How often do you fall in love?" Negotiators fall in love the most frequently of the four basic types. If adventure is essential to the Explorer, and loyalty is imperative to the Builder, and sex is vital to the Director, romance is critical to the Negotiator. So these men and women plan romantic evenings and weekends; send romantic cards and e-mails; express their love regularly with hugs and endearing words; and hope for clear, regular signals that their devotion is returned.

Sex, for the Negotiator, is another essential part of a partnership. But sex must be mixed with romance, communication and emotional intimacy. Sex is not casual to Negotiators; if they engage in it, they expect a meaningful relationship to develop.

Yet even sex with a familiar partner can have some minefields for this personality type. Negotiators have a rich fantasy life. And they can idealize a lover, regarding him or her as perfect. Then, as they get to know their mate, they can begin to see their partner's imperfections and become disillusioned. Their fantasy is tarnished. And their negative feelings can contaminate not only their life in the bedroom but the entire relationship. For Negotiators, sex and love are utterly intertwined.

Negotiators are far more complex than they often seem. While they deeply value harmony and closeness with a partner—and spend

an inordinate amount of time and energy building a profound connection with this soul mate—they don't like clingy or demanding
people. So if a partner begins to require an inappropriate amount of
their attention, they can feel burdened and pressured to provide more
love, as well as eager to be free. Negotiators expect an eternal union,
but each partner must respect the other's needs.

Different types of Negotiators have different needs and desires,
of course. NEGOTIATOR/Explorers are lovers through and through.
As Negotiators, they seek a bottomless, spiritual connection with a
"true love." As Explorers, they are particularly attracted to individuals
who are daring, playful and adventurous. But both Negotiators and
Explorers can be chameleons—so charming, personable and eager to
please that they can be difficult to get to know. The same can be said
about NEGOTIATOR/Explorers. They might not know who they
are either—because they have so many angles to their personality. Yet
these men and women bring much to a relationship, regularly kin-
dling imaginative and creative conversations and experiences that are
interesting and fun.

NEGOTIATOR/Builders are superb at handling people. They
are sympathetic and tactful, and intuitively know what others are
thinking and feeling. With their command of language, their enthusiasm and their acute social sensitivity, they can be warm and charming
on dates. But NEGOTIATOR/Builders are the most pragmatic of the
Negotiator types. To them, love must be embedded in a stable, long-
term relationship. They are also traditional, believing that this emotional connection should culminate in a march down the aisle. They
want a partner who makes plans, adheres to schedules and has a fixed
moral view.

NEGOTIATOR/Directors, like all Negotiators, seek timeless
love with the "perfect" partner. They intend to bond in a spiritual sense,
communicate their deepest inner feelings and enjoy an intense, meaningful and lasting relationship. But NEGOTIATOR/Directors can be
straightforward, tough-minded and analytical. And despite their poise
in social situations, they are the most likely of the Negotiators to

engage in focused, even competitive conversations, intensely exploring ideas, insights or spiritual beliefs.

Mark Twain once said, "Put all thine eggs in one basket and— watch that basket." Negotiators do. Marriage is important to most of them. But the social pledge of matrimony is far less sacred to many Negotiators than the deep personal commitment they privately make to their partner. To Negotiators, this personal vow is the true marriage vow; external rituals do not sanctify their devotion. Many even write their own vows to create a nontraditional wedding ceremony.

In fact, I have come to believe that Negotiators expect of their partnerships something the other three types often don't seek, don't understand and don't even appreciate: absolute and unconditional love. If the Explorer expresses more *Ludus,* playful love; and the Builder seeks more *Pragma,* love based on compatibility and common sense; and the Director displays more *Storge*, a deep feeling of special friendship for a partner that lacks overt displays of emotion; the Negotiator expresses more *Agape*—the Greek word for gentle, unselfish, all-giving, unconditional love.

As a result, Negotiators tend to stay in a relationship far too long. Their sense of responsibility and compassion, their patience and their drive to create this unconditional connection override their disappointment in a poor match.

Still, if the Negotiator comes to believe his or her partner is not a soul mate, and cannot grow to become a soul mate, he or she eventually departs to search for perfect love.

"Clicking"

So where are we?

We all have personalities. Roughly 50 percent of the variations in your personality arise from your biology; these are your inherited traits of temperament. We each express a primary and secondary personality type. And your primary style of thinking and behaving guides you toward certain mates. Explorers are initially drawn to other Explorers, and Builders tend to "click" with other Builders. Directors, on the other hand, gravitate foremost to Negotiators, while Negotiators are drawn to Directors. Moreover, each type seeks different things in a partnership and courts in different ways.

But there is more to "mate choice" than your temperament. So before I examine how these basic personality types actually get along with one another in a long-term partnership, I must consider some of the other forces that drive us to choose "him" or "her."

CHAPTER 9

Thunder opened chasms of light
In your dark eyes,
In the night,
On the water.
And in me the storm was born.

—LUCHA CORPI

FALLING IN LOVE:

The Funnel

"It may not happen in the first instant, but within ten minutes of meeting a man, a woman has a clear idea of who he is, or at least who he might be for her, and her heart of hearts has already told her whether or not she's going to fall in love with him." Orhan Pamuk, the Turkish Nobel laureate in literature in 2006, knew that women can feel an almost instant attraction to another. Men can, too.

Love at first sight doesn't happen to everyone. In one study of almost a hundred men and women, it had occurred to some 10 percent of them. Yet we can all walk into a room full of people we don't know and pick out one or more individuals to whom we feel attracted. It takes less than a second to make this judgment.

Each personality type is especially attracted to certain others, of course; temperament always plays a role in whom you choose. But other forces also contribute to that spark of romantic ardor. Your childhood experiences, the things you need and want from a beloved, a potential partner's looks and voice—these and many other subtle forces can ignite or quash feelings of romantic passion. Love operates as a kind of funnel. As you get to know people, you get to know more and more about them, and gradually or rapidly you weed out all—until you find "the one."

This process of *mate choice,* as scientists call it, often begins with something beyond our control: timing.

A Time to Reap

There's a "time to reap," as the Bible says. We tend to fall in love when we are ready. The near-perfect partner can be standing by the door as you walk into a room, but you might not notice him or her unless you are ready to start a new romance.

We don't always know when we are ready, though; romantic love can strike when you least expect it. But there are certain times when all four personality types are particularly vulnerable to this bliss: during a life passage, a transition, a period of change. If you've just moved to a new city, entered college or started a new job; if you've recently divorced, recovered from a broken heart or are "between things" in some other important way, you are more disposed to falling in love.

This was true of Patrick, the man whose wedding I attended a few months after he signed up with Chemistry.com. Patrick was adrift, uprooted from his home, his career and his community in New Orleans. Hurricane Katrina had stolen all. His bride, Suzanne, must have been ready, too, because she had also joined Chemistry.com to look for a long-term mate.

This susceptibility probably has a chemical component. Transitions produce stress and strong emotions. Anxiety; panic; terror; fury; jealousy; elation: all activate chemical systems in the brain that can escalate feelings of romantic passion. So as you struggle through one of life's upheavals, the circuits in your brain for general arousal, energy, focus and motivation can push you closer to the threshold for falling in love.

Wartime unquestionably stirs one's romantic passion. In 2006, Nadia Habib, an Iraqi wedding planner in Baghdad, told a *New York Times* reporter that she had recently transported an extraordinary number of fifteen-layer wedding cakes with buttercream frosting through the checkpoints and convoys of American humvees that were snarling traffic in the 110-degree heat. Explaining the unusually high number of weddings, she said, "The sense of danger intensifies people's feelings for one another."

Still, I suspect that each personality type falls in love during times of stress or joy for slightly different reasons. Explorers like change; novel, unpredictable situations often thrill them. In this invigorated state, they may become even more flexible and optimistic than usual, noticing individuals they would have overlooked in a less stimulating environment. Builders, on the other hand, crave order and stability. So as Builders become more and more agitated under stress, they may seek a partner as a haven in the storm. Directors are saviors; they want to be useful, to be needed. So as they help those around them during a crisis, they may fall into the arms of a beloved. As Negotiators become increasingly anxious and emotional under stress, they probably employ their standard "tend and befriend" response, the drive to build deep connections with others. And as Negotiators reach out to connect, they may stumble upon "the one."

Proximity

"Ah, I have picked up magic in her nearness." So wrote poet Ezra Pound. Proximity also contributes to romantic passion in all four personality types. At work; in class; at the gym; in the neighborhood: you tend to fall in love with those who are nearby. In a recent study of a hundred young American men and women, sixty-three said they had fallen in love with someone because he or she was "around."

The power of proximity was vividly expressed in an e-mail I recently received from a man in Canada, Sandy. I suspect Sandy is an Explorer. He wrote: "I used to think I would never marry. It just seemed like too much trouble. I had a lot of girlfriends. But I could never really see any reason to tie the knot. But I kept seeing a woman who lived across the street; we would make idle chit-chat as we came and went. She wasn't my type at all. I never would have approached her at a party. But I became fond of her. Eventually I fell in love with her. Now we have been married for 32 years and I am still deeply in love with her. She is warm and loving and a great mother to our three children. I'm just grateful that she lived next door; she made my dreams come true."

Data indicate that the more you interact with someone, the more you are likely to regard him or her as good-looking, interesting, smart and similar to yourself. The better you like them, too. But one's passion for those who are "around" is probably not random. People with similar interests and backgrounds, and of similar ages and economic means, tend to congregate in the same places. My Personality Type Study showed that Explorers and Negotiators prefer to live in a big city. Builders like the suburbs. And Directors choose the mountains.

So if you're an Explorer or Builder, head for where you want to live, then seek a mate; your ideal partner will be there too. If you're a Director, you'll have to find your Negotiator where the action is, then transport your family to a mountaintop—or at least take them there on vacations. And if you're a Negotiator, live on the high peaks or anywhere else; Directors have the focus and determination to find you.

Familiarity and Barriers

All four types are also likely to be attracted to individuals who are somewhat familiar. *Repeated exposure,* as psychologists call familiarity, seems to lessen dread and heighten interest.

But I suspect that Builders are the most likely to wed the girl or boy they've known since childhood, because Builders are particularly comfortable with what and whom they know. Explorers, on the other hand, are probably the most inclined of the four types to seek a mate outside their circle of friends and kin, perhaps even someone from another country who speaks a foreign tongue. Because Negotiators are also mentally flexible, as well as particularly compassionate and interested in people of all kinds, this type may also marry more frequently outside their familiar realm. And Directors want a mind mate, so familiarity will not be a drawing card for them either.

But even Builders are not likely to fall in love with someone they know extremely well; *too much* familiarity is a romantic turnoff, as life on an Israeli kibbutz has shown.

In the heyday of these collective farms, all children grew up together, eating, sleeping and bathing in their common "Children's House." By age twelve, however, boys and girls became tense with one another and began to avoid any sort of physical contact with others in their group. Then, as they matured, they began to regard one another as brothers and sisters. In one study of an Israeli kibbutz, no two members who had shared a common cradle when below the age of six married each other. As a result, scientists have concluded that at some point in childhood, probably between ages three and six, boys and girls who are exceedingly familiar with one another lose forever the ability to fall in love with these individuals.

Mystery is indeed an aphrodisiac. So even Builders who marry their childhood sweetheart probably didn't know that person intimately when very young.

Barriers also spur romantic love. If you believe someone is unattainable, you tend to like him or her even more—what psychologists call the *Romeo and Juliet effect,* and what I call *frustration attraction.* Some of the brain circuitry of this phenomenon is known. When foiled, the brain's system for *wanting* just keeps pumping, urging you on with focus, energy and incentive.

Mirror, Mirror: Looks Count

So, let's say the timing is right: you moved into town three months ago. You have felt lonely yet exhilarated for weeks; your emotions are running high. And you have come to a party with some friends from

the office. You are standing with acquaintances when suddenly someone new joins your cluster. You have seen him or her "around," you
can't remember where. The newcomer is slightly familiar yet mysterious. And you are ready to fall in love. What happens first?

You look. Surreptitiously you scan this novel creature with your
eyes—even before you are introduced.

Looks count. The male brain is particularly well built for sizing
up potential mates by looking at them, as my colleagues and I noticed
when we put men and women into a brain scanner and collected data
as these subjects looked at photos of their beloved. Our male subjects
showed more activity than females in regions of the brain associated
with *visual* processing, particularly of the face.[1]

No wonder women go to such extraordinary lengths to adver-
tise themselves visually. This heightened male sensitivity to looks may
also help explain why men fall in love faster than women.

But just about everyone wants a partner who's good-looking.
Studies show that we tend to regard people we find attractive as confident, smart, stable, friendly, generous, self-effacing, sensitive, finan-
cially secure, healthy and popular—traits most people seek in a mate.
And we are particularly drawn to those with symmetrical features.

Chimpanzees share our taste; they also prefer symmetrical features in a mating partner. In fact, symmetry is the essence of beauty
to many mammals and birds, even insects. Female scorpion flies prefer males with uniform wings. Peahens prefer peacocks with symmetrical tail feathers. Monkeys favor consorts with symmetrical teeth.
Even two-month-old human infants gaze longer at more symmetrical
faces.

Why? Good-looking people aren't necessarily nicer, smarter or
more compassionate than the rest of us. But symmetry sends a special message to the unconscious core of the brain: good genes and
good health. It's an honest message, too. Individuals with well-
proportioned, symmetrical features have been able to avoid accidents
and combat the bacteria and viruses that produce irregularities. In
short, symmetrical men and women have sturdy immune systems.

So the human taste for a symmetrical partner is an adaptive mechanism that evolved to motivate us—as it does many other creatures—to select biologically "fit" mating partners. Fortunately, however, both men and women also tend to be most attracted to people who share their own *degree* of good looks.

But each personality type probably has particular preferences. Explorers have more expressive faces, so these men and women may be more likely to be drawn to an animated, energetic visage. Builders, on the other hand, may be more drawn to those who look conservative—women with less makeup and men with a more traditional haircut. Directors clearly seek those with signs of high estrogen, such as round faces and full lips, "baby faces."[2] And Negotiators are drawn to those with broad, angular faces and other facial signs of testosterone.[3] Negotiators may also like a less expressive face, a "poker face," because testosterone tends to reduce one's facial expression.

Back to the party.

You've looked. You like this stranger's face. What next?

Shape Counts

The body.

As deftly as possible, you check out this mysterious person's frame. Men tend to be more attracted to a woman with a beautiful body and less inviting face than to one with a gorgeous face and an unappealing figure. Height is not an issue for men; they like women of any reasonable height, although as a rule they prefer shorter women. And most men prefer voluptuous, curvy women to thin ones.

In fact, studies show that men in the United States, Britain, Germany, Australia, India, Uganda and several other countries are particularly attracted to women whose waist circumference is about 70 percent of their hip measurement—probably because women with this *waist-to-hip ratio* are more likely to bear healthy young.[4] Men in ancient Europe, Asia, the Americas and Africa also preferred this particular shape. We know this because thousands of years ago they were carving statues of women with these proportions.

Women often downplay the importance of bodily appearance in a potential sweetheart. When they are hooked up to lie detectors, however, they admit that physical appearance plays a big role in their feelings of initial attraction. Women are drawn to men with a developed chest, broad shoulders, narrow hips and legs, and small buttocks. Women also like tall men (an indication of testosterone), perhaps because tall men tend to get better jobs and make more money and are more likely to become powerful in business or politics. Tall men look more intimidating, too, and thus potentially more protective.

Women are also drawn to men who stand or sit in a carefree pose, a sign of dominance; to those who are self-confident, assertive and smart; and to those with a deeper voice, another sign of testosterone. And women are particularly fond of men who are coordinated, strong and courageous, as well as men with money, status, education and resources. In fact, everywhere in the world *big men,* as they are called in hunting/gathering societies, are prized—for the obvious reason that rich men can provide.

But, once again, each personality type has some slightly different preferences. Explorers have a relaxed, fluid, energetic way of moving that probably particularly appeals to other Explorers. You may have noticed that Barack Obama moves in the smooth, graceful style of the Explorer.

Builders send a different signal. When I briefly met a group of Builder men and women at a singles party, I was struck by the number of men wearing a suit and tie—thus advertising their business savvy. They also held themselves in a formal pose, sitting or standing as if planted rather than about to move. Female Builders were more formal, too.

Both male and female Directors most likely gravitate to more feminine bodily features—the soft, unthreatening look.

And Negotiators (both men and women) are probably particularly attracted to tall partners with slim hips. I know some Negotiator women who will go out *only* with tall men. They are dancing to nature's rhythm—searching for their counterpart, high-testosterone Director men.

The Voice: Grooming Talk

So there you are, standing with your friends at the party. You have appraised the full figure of this still mysterious person standing near you. And you still like what you see. Now it's time to talk, so you strike up a conversation about the food.

Instantly you are struck by the *way* this person talks. The voice says volumes about one's background, education and intentions. If the accent isn't "right," it's over. The music of the voice is important too. If he quacks like a duck, if she sounds too chirpy or she growls, you're repulsed. People regard rapid talkers as better educated, and those with more voice inflection as more interesting. In fact, if you have a warm, full, impressive voice, people actually regard you as better looking. Indeed, you probably are. People with voices that others have rated as "attractive" are also more physically symmetrical.

But let's say that in the case of this mysterious stranger, you like both the accent and the voice inflection. Now for the words. Words make up only about 10 percent of the vocal signal people broadcast. Their vocal tones make up another 35 percent of their message and their body postures and gestures comprise another 55 percent.

Even so, words are tiny bombs. If you are an Explorer and you hear too many Builder words, like *proper, moral* and *schedule,* you will probably speed off to get a drink—alone. If you are a Builder, you might perk up, even focus a bit harder on this new member of the group. As my Word Type Study showed, each personality type is far more attracted to some words than to others.

So let's say the exotic person you are getting to know at the party has the high sweet voice that appeals to men or the deep confident voice that women like. He or she is standing and moving in ways you find attractive, as well as using words that appeal to your core nature as an Explorer, Builder, Director or Negotiator. More hurdles have been cleared.

Suddenly the music starts. Seized with curiosity, you ask the mysterious newcomer to dance. And he or she accepts.

Rhythm's Message

Almost immediately, waves of anxiety hit you in the stomach. You think: what if this enchanting creature can't keep the beat? Even worse: what if you maul him or her with your own fat feet? Why do we feel so nervous when we first dance with a prospective sweetheart?

Because the way we move says reams about who we are, as anthropologists Robert Trivers and Lee Cronk have established. These scientists taped laser reflectors to several moving body parts of young Jamaican men and women, including their shoulders, elbows, wrists, hips, knees and ankles. Then they used cameras to record the Jamaicans' movements as they danced. Last, Trivers and Cronk displayed *only* the moving lights on a computer screen and asked other Jamaicans to pick out the best dancers. The observers easily identified the most coordinated participants.

But these rhythmic men and women were more than just good at keeping time; they were also more physically symmetrical. So as coordinated men and women hold and move their bodies in rhythmic ways, they are actually showing off their strong immune systems and genetic strength.

Odor Prints

Happily, you have weathered this short whirl around the floor with the newcomer, and all toes are still intact. You also achieved some rhythmic synchrony, a good sign because studies show that brain waves get in sync when you keep the same beat. Even better, as you spun around the floor your partner gazed at you for a long moment. And this eye language, known as the *copulatory gaze,* sent a shiver down your spine.

Equally important, while you moved rhythmically together you caught a whiff of him or her—and found it pleasing. This is essential. Tobacco, liquor, spicy food, perfume or cologne, lingering odors from the gym—dozens of scents compose your *odor print.* And smells make a beeline to the brain's primitive centers for memory and emotion. Here they detonate—causing immediate sensations of attraction or repulsion.

Moreover, these aromatic missiles say vital things about a person's well-being. Women and men with what others consider sexy body odors are more likely to have symmetrical faces, signs of a strong immune system and good health.

It's a good beginning.

Courtship Talk: Similar Values

So you leave the dance floor and your group and head for the bar to get a drink—together. Now begins a serious stage in the wooing process: searching for compatibility.

As you recall, Explorers like adventure and activity, so if you are one of these sensation seekers you are likely to strike up a conversation about something you did recently. Builders, ever precise and literal, like facts and details. They also tend to be social. So a Builder is more likely to begin with casual chat about a recent political, sports or social event, or someone they know at the party. The Director may dive into an intellectual conversation on an important, perhaps esoteric topic. And a Negotiator may embark on how he or she feels about a broad, often philosophical issue.

Regardless of how this conversation begins, an Explorer is soon likely to ask you what you *do;* a Builder will be more interested in finding out whom you *know;* a Director will want to learn what you *think;* and a Negotiator will inquire about how you *feel* about some issue or event. It looks casual, but it's not. Nothing is casual at the beginning of a courtship. You are hanging on every word, every phrase, every detail of the conversation. You are overweighing these details, too. With so little information about him or her, every crumb of data counts.

Foremost, you are searching for your new partner's values. And values must be shared.

If you believe that captains of industry should be free to "develop" the world's natural resources, while the newcomer eats tofu and hugs trees to protect the planet; if you go to church every morning on your way to work, while the stranger shares H. L. Mencken's view that religion is a "high and ghostly matter" that can't be fathomed; if you like to spend your Saturdays with your buddies on a party yacht, while this mysterious person likes to go to art museums and play the violin—if you and your would-be partner suddenly realize that you hold very different core values, both of you will soon drift to opposite ends of the buffet line. The romance is over.

Men and women are attracted to individuals who share their standards, ethics, morals and ideals. This is why psychologists (and some dating services) maintain that similarity attracts. When it comes to *values,* similarity does attract.

All four temperament types also gravitate to those who come from their own socioeconomic and ethnic background; those with the same general level of education and intelligence; those with similar political and religious views; those with similar social goals; those with the same degree of financial stability; those with similar social and communication skills; and those with a similar sense of humor. In all these ways, like attracts like—what anthropologists call *positive assortative mating* or *fitness matching.*

It sounds like a lot of work to weigh all these criteria during one's first conversation with a potential life partner. But psychologists believe we make up our minds about whether an individual could be an appropriate mate within the first three minutes of talking to him or her. With no contradictory or modulating information about this unfamiliar person, we weigh these first impressions heavily, dramatically influencing how the relationship will proceed.

But let's say that in this case, you discover that you do happen to share many values: the newcomer hates the same politicians you hate; celebrates Christmas, Ramadan or Hanukkah the way you do; and eats only vegan or loves steaks and fries, just like you. As the evening progresses, you discover that this appealing individual also shares many of your interests: you both like to ski, go on long walks with your dog, play cards and read the same newspaper.

So far, so good.

But another vital benchmark lies ahead: can this person give you what you need?

Exchanging Needs / Providing Roles

"Tell me whom you love and I will tell you who you are and, more especially, who you want to be." Psychoanalyst Theodor Reik believed that men and women choose mates who satisfy an important need, including the qualities they themselves lack.

John W. Warner IV, the son of the Republican senator from Virginia, appears to have found what he needs in Shannon Ford Hamm, as she has in him. The *New York Times* recently reported that after four years as "hellmaster" of his fraternity at the University of Virginia, John turned to racing cars. "It's a thrill going around the track six inches from the wall at 180 miles an hour," he said hours before his nuptials. But he grew tired of the speedway, of playing pool and of squiring supermodels to after-hours clubs. He matured—and fell in love with Shannon, an educator from Minnesota who taught first grade at the Spence School in New York.

As one guest remarked on their wedding day in 2007, "These two people are just so meant for each other. She is enormously patient, and he's enormously impatient and it really goes together."

Social exchange theory holds that men and women are attracted to those who can provide them with the resources they seek in exchange for the assets they can provide. The exchange may be of any kind—but it must be "fair." Both partners must believe they are acquiring the right rewards for the right price.

Perhaps she trades her superior education as a doctor for his upper-class pedigree. Some trade their tallness for the other's shortness, consciously or unconsciously hoping to produce children of average height. Submissive individuals often gravitate to those who are dominant. In a classic study of some ten thousand men and women in thirty-seven countries, men were more likely to seek women who were young and beautiful—women who could bear them healthy offspring. And women more regularly sought men with status, education, money or other tangible goods and services they needed to rear their young.

So as you talk at the bar with this engaging person about what each of you does, whom you know, what you think and how you feel, you are also sizing up your potential partner's assets—what each of you can bring to the relationship. If either of you discovers that the other is already involved in a long-term relationship, is about to move to Timbuktu for the next five years, is flat broke or has some other nonnegotiable trait, you will lose romantic interest instantly.

The quandary comes, of course, when you perceive that this intriguing person can meet some of your deepest needs, but not others.

In fact, I suspect that all of us have a hierarchy of needs; some we are willing to forgo, while others absolutely must be filled. Moreover, no two people's nonnegotiable needs are exactly the same. I've seen men and women who are willing to overlook a lousy sex life with a partner who can give them the intellectual environment they crave. Others will tolerate alcoholism or adultery if they can get the money, influence or connections they feel they can't live without.

I remember a beautiful young assistant of mine once telling me matter-of-factly why she had decided not to marry a particular man. She said, "I'm smarter, richer and better looking than he is; I need an equal." I agreed.

This *need complementarity* is often essential to a strong relationship. If the costs outweigh the benefits, you are not likely to pursue the match.

But as you stand there at the bar, talking and listening, you are beginning to feel that this charming individual might be able to provide what you require. This exotic creature comes from your general economic, social and educational background; shares your values; is single and available; and also appears to have a good job, live in the "right" part of town, know some interesting people, even do some of the things you like to do. Moreover, this possible life partner is passably good-looking and amusing, even able to keep your beat on the dance floor.

You stand a little closer to him or her.

The Roles We Play

Now for another serious consideration: your role in the relationship.

As you talk of politics, golf, poetry or the weather, you secretly wonder if you could really play the roles required of a partnership with this individual. If you suddenly find that this intoxicating person is a sword swallower in a traveling circus or in some other way lives a life you can't imagine joining, it's over. If this individual is an academic, financier or artist instead, and you think you could "fit in" as this person's spouse, your romantic interest grows even more.

I recall the moment I understood the importance of one's role in a relationship. I was visiting a girlfriend after making a speech in upstate New York when her mother pulled me aside to whisper, "You don't marry a person; you marry a lifestyle." It's true. Your daily life with an Explorer is likely to be very different from your routine with a Builder, Director or Negotiator. But I think you marry a person *and* a lifestyle. The role you will play in a relationship is only one of many vital make-or-break points in a budding partnership.

So let's say that you and your new friend have established the roles that each is likely to play in this potential partnership, and you're feeling even more compatible. Ready to take the leap?

No, because here comes the most subtle and dangerous obstacle of all: your childhood and your other past experiences.

Love Maps

Childhood is "damned serious business," said the children's book author and illustrator Maurice Sendak. He was right. As you grow up, you begin to build a largely unconscious list of traits you will eventually seek in "him" or "her." Psychologist Marcel Zentner calls this complex psychological chart your *Ideal Mate Personality Concept;* I call it your *love map.*

Everybody's love map is unique. Your mother's zest; your father's way with words; how your parents, siblings and other kin express love and handle money; what your childhood friends think is funny; what your peers admire; how your teachers viewed honor, justice and kindness; what your community cares about: these and hundreds of other forces and events converge to create a unique constellation of interests, values, beliefs and behavioral idiosyncrasies you carry in your head. Then this largely unconscious psychological map guides you as you navigate the sea of mating opportunities and finally settle on "the one."

This is one reason why it is difficult to introduce single friends to one another; we don't know their love maps. It also helps to explain why some couples remain together even though none of their friends can fathom why.

Take my friend Mary. Mary grew up with an alcoholic father, a lawyer by profession. As a child, she became used to her dad's unpredictability. She never knew when he would be attentive and compassionate, when surly, when violent. She learned to expect drama every day and night. She was her mother's accomplice, too, helping to hide the truth from neighbors, family and friends. Mary resolved she would never marry an alcoholic like Dad. She didn't. Whom did Mary marry? A sober, yet unpredictable, undependable artist. Mary's love map had given her a taste for the capriciousness and excitement she had learned to expect as a child.

Building Your Love Map

Psychologists have offered many theories about the primary forces that contribute to your love map.

Freudians maintain that women tend to fall in love with men like their father, whereas men are attracted to women much like their mother. They believe we are attempting to reunite with our first true love: the parent of the opposite sex.[5] Others believe we choose a mate who is similar to the parent with whom we have unresolved issues, a partner who suffered traumas similar to our own in childhood, a partner who will provide the type of infant attachment we made with our own mother, be it secure, ambivalent or avoidant, or a partner who will reflect the values and interests of our childhood friends.[6]

I think *all* the forces mentioned above—and many more—affect your love map. Yet each of us is influenced by certain experiences more than others, creating our own unique tapestry of requirements for loving. Moreover, these love maps change over time. As you ride the waves of life—and experience a few romantic disasters—you trim and deepen your image of the ideal mate.

Positive Illusions

Interestingly, women have clearer, more incisive love maps than men. And women are better than men at picking a partner who fits within their love map. This feminine skill pays off, too. People are happier in a long-term partnership when their mate has many of the traits that fit within their ideal.

Nevertheless, even a good love map fit is not as effective as your *positive illusions*. I say this because the happiness you will find in a relationship depends less on whether this individual actually fits within your love map than on whether you *think* this individual fits within your love map. Known as *love blindness*, this may be one reason why self-deception evolved—to keep us happy with a romantic partner.

"Love Begets Love"

It's getting late. But there you are, still chatting with this appealing person. The two of you have talked and flirted, danced and shared a drink, told each other what you did last summer, exchanged some thoughts about world affairs, established whom you both know in town, and discovered that you like the same music, art and sports. Moreover, this person you are talking to fits your concept of the ideal mate. You wonder: will this be our first—or last—conversation together?

Suddenly something happens—*and you know.* Perhaps it is the way he asks you for your phone number a second time. Or the way she smiles at you when you say it's time to go. But in an instant you realize this person is seriously interested in you. And that thought starts a fire in your brain. "Love begets love," wrote poet Theodore Roethke. Men and women tend to fall in love with people who are in love with them.[7]

You say good night. Nature has played her primordial melody, and this new person in your life has slipped easily through your funnel of cultural and biological criteria. You don't know it, but even your personality types are in sync. And as you turn away, you feel the first tingling ecstasy of intense romantic love.

CHAPTER 10

As there are as many minds as there are heads, so
there are as many kinds of love as there are hearts.

— LEO TOLSTOY

THE ONE:

How We Love

Chestnut trees, kapok trees, breadnut trees, palm trees, mahogany trees, chicle trees—trees, trees and more trees bathe lowland Guatemala in green. Howler monkeys and parrots chatter in this universe of leaves. And far below prowl anteaters, peccaries, tapirs and jaguars. Gone are the sweeping boulevards and gracious palaces that once reigned here, swallowed by roots and vines. But a majestic temple has been restored: the tomb of King Hasaw Chan K'awil, who was buried beneath this citadel of stone around AD 720. He had stood over six feet tall, lived into his eighties and was the grandest sun king of the grandest city, Tikal, of the grandest New World empire, the Maya.

Mayan inscriptions proclaim that Hasaw Chan K'awil adored his wife. He also built a temple in her honor—facing his. And twice annually, on the spring and autumn equinox, the sun rises behind his royal dwelling to cast its shadow directly across her temple. Then, as the sun sets, it spreads the shadow of her temple directly across his tomb. With this celestial device, these lovers still touch some thirteen hundred years after death.

The "Best" Match

What kind of man was Hasaw Chan K'awil? What kind of marriage did he make? Perhaps he was an Explorer who fell in love with another Explorer like himself. Perhaps he was a Builder married to a Builder. Or a Director who wed a Negotiator. He could have made any kind of match. But he was happy in his marriage.

People often ask me which is the best match, which the worst. I think each type of partnership is simply different. An Explorer wed to another Explorer will have different joys and sorrows than an Explorer married to a Builder, Director or Negotiator, for example.

Moreover, your secondary type always plays a role in how you love, as was recently described to me quite vividly. I had just made a speech about these four basic personality types, and a woman in the third row raised her hand and said, "I can really see how this all plays out in my house. I am an Explorer, and my husband must be an Explorer, too. But my secondary type is definitely the Negotiator, while his must be a Builder. Sometimes we get into a really stupid fight. That's probably the Negotiator fighting with the Builder. Then we just drop the whole thing and go hiking." In this case the free-spirited Negotiator and traditional Builder couldn't understand each other. So both reverted to their primary disposition—the Explorer—to find union.

We use many aspects of our temperament and our experience as we make a partnership. Moreover, relationships grow, change and often mature. In fact, some psychologists maintain that one's real personality plays only a minor role in the first eighteen months of a budding romantic match. Only after you settle in with "him" or "her" do the more deep-seated aspects of your temperament and character emerge. Even then, it sometimes takes a major crisis to realize how stubborn, flexible or compassionate a partner really is.

So although my Mate Choice Study (see the appendix) indicates that we are naturally drawn to some types more than others—and I suspect these matches are also more harmonious—different pairings of the four personality types yield ten *basic* types of partnerships. All can work, but each will have its ups and downs. I'll explore all ten. But please keep in mind that each of us is a combination of all four basic personality types, and that the secondary aspects of your temperament and character affect who you are and how you love.

Explorer-Explorer Match

One Tuesday morning some years ago, my Explorer boyfriend looked at me over breakfast and said, "Ever been to Warsaw?"

"No," I replied.

"Wanna go?" he asked.

"Absolutely," I said.

He had just seen an ad in the *New York Times* for cheap round-trip tickets to Warsaw. Three days later we stepped off the plane in Poland and headed for the Warsaw Uprising Museum, then dined in the Old Town. Havana, Athens, Kangerlussuag, Reykjavik; Shiraz; Bratislava; Istanbul; Cairo; Dubrovnik; La Paz: I have wandered through these and many other places with this Explorer. With him or other Explorers I have also read poetry and stories, explored the world's great museums, seen hundreds of plays, symphonies and operas, and hiked the mountains and back streets of many exotic places. I have been fortunate; over the years I have found several Explorers eager to go adventuring with me.

When two Explorers pair up, they do things together. Both are curious. Both have many interests, love to learn, and seek novelty, excitement and adventure. And both are likely to be impulsive, high-energy, flexible, daring, irreverent, resourceful and very interested in sex. So two Explorers can have great fun. Each has found someone who will play as hard and fast and creatively as they had always dreamed a partner would.

Explorers also tend to be spontaneously generous. My Explorer used to give me stamps and paper clips for Christmas, then surprise me with an extravagant present on some otherwise ordinary day. Two Explorers aren't likely to argue over money, either; most share the view that money should be used to fulfill their dreams. Nor are they likely to argue over life's little chores. How to mop the floor or take out the garbage is likely to be immaterial to both.

Explorers are flexible. They don't adhere to schedules or pre-scribed ways of doing things unless necessary. So neither will be fussy about rules. Explorers are optimists, too. Both partners are likely to look on the bright side of life, as well as minimize serious matters. Their light touch, their adaptability and their mutual irrev-erence enable both to keep their spirits up when their life together is strained.

But this match is not for the faint of heart. With a shared willing-ness to tolerate risks, two Explorers can find themselves in magical—or disastrous—situations. I remember one fiasco particularly clearly, al-though it happened more than thirty years ago. One Saturday morning in January, my (then) Explorer boyfriend bought a jeep in Long Island, New York. We decided to drive it along a deserted beach around midnight. For a thrill, he drove close to the waves, and in sec-onds the car sank in the sand. The tide was rising. As he revved the en-gine and spun the wheels, the jeep sank even more. Finally I decided to hitchhike to the nearest town, find the police department and mar-shal a host of cops, sleepy citizens and tow trucks in ten-degree weather to rescue him. The excitement wasn't worth the strain.

Despite their mutual attraction, a pair of Explorers are bound to have problems. What they gain in intellectual and/or physical adven-ture, for example, they can lose in intimacy. Explorers are wildly inde-pendent; they must have autonomy, even from a mate. Explorers aren't generally introspective, either. So two Explorers will often shun diffi-cult discussions, leaving serious family issues unresolved.

Explorers also tend to be flirtatious. And they can stumble into romantic situations or even seek sexual involvements with men or women other than their partners. They are also prone to addiction—to drugs, gambling, alcohol, sex or work. And two Explorers can encourage each other's destructive habits and spend too much time together in bars or casinos. More insidious, two Explorers can burn out—exhausting each other intellectually, physically and emotionally. They can feel as if they've done it all and then hanker to move on as the novelty wears off.

But when this match works, it really works—as I discovered when I did a study of five hundred couples for *Oprah* magazine in 2007. In this investigation I correlated various types of matches with scores on a happiness scale. Some of my questions measuring happiness were "Does your partner satisfy your deepest psychological needs?" "How often do you feel 'in love' with your partner?" and "Do you get enough affection from your partner?" Couples were largely men and women married for fifteen to nineteen years. Among my findings, Explorers married to other Explorers were particularly happy.

In Keats's poem "Ode on a Grecian Urn," the male lover is painted on the urn in "mad pursuit" of his beloved, "For ever panting and for ever young." Keats captured the essence of the happy Explorer-Explorer match.

Explorer-Builder Match

Explorers and Builders are not typically attracted to each other; this is not a natural partnership. In fact, my Mate Choice Study showed that male Explorers were mildly *unattracted* to female Builders; and male Builders were slightly *unattracted* to female Explorers (see the appendix). This is entirely understandable. Explorers and Builders are polar opposites, even biologically: dopamine suppresses serotonin channels, and serotonin suppresses dopamine circuits. So although this match can be dramatically effective, it's often challenging.

Sometimes these opposites attract. The calm, controlled, cautious Builder, for example, may be unconsciously drawn to the energetic, impulsive, risk-taking Explorer, eager to enjoy the Explorer's vitality and sense of adventure. The Builder can also benefit from the Explorer's playful sexuality. And the Builder can use his or her nest-making and protective instincts to care for the Explorer. Moreover, the Builder is likely to listen attentively to the Explorer—who often likes to talk about his or her ideas.

Meanwhile, the novelty-seeking, irreverent, creative Explorer can be attracted to the stable, loyal, traditional and social Builder because he or she wants family, home and community—the roots the Builder can provide. The Explorer may also admire the Builder's rectitude, patience, trustworthiness, industry and managerial skills. The Explorer gets a chance to amuse and tease the Builder. And since Explorers often want to lead the way, in the Builder they may have a partner who wants to organize their adventures and come along.

So an Explorer and Builder can satisfy some of each other's basic needs, as well as provide their children with both adventure and security.

But this match is likely to face problems.

Builders are not casual about courtship, sexuality or their desire to start a family. They tend to be moral, responsible and conventional. Builders have clear values, and they stick to them. Hence, the Builder may try to push the Explorer to settle into a steady job and daily routine. This can cause tension.

A big issue is the Explorer's need for novelty versus the Builder's love of familiarity and schedules. This difference became apparent to me one morning in a London hotel lobby. An Explorer friend of mine and her Builder husband were discussing their jogging route. It was their first day in London. The Explorer suggested they lope through the deserted city streets, past Buckingham Palace and other sites that neither had ever seen. But the Builder wanted to gauge the exact time and distance of his run, so he recommended they jog on the treadmills in the hotel basement. Each scowled at the other. Eventually the Explorer sped through London's byways while the Builder descended to the gym.

Time may bring other problems—particularly regarding rules.

Explorers don't respect rules; rules stifle their autonomy and spontaneity. So the Explorer can become irritated by the Builder's compliance with regulations and traditions. Yet Builders want to behave
sensibly; for them, the respect of others in the community is essential.
And they can view the Explorer's disregard for rules and traditions as
reckless and disrespectful. But trying to get an Explorer to conform to
prescribed standards is like using a net to catch the wind.

Their differences in temperament can cause friction in more
subtle ways. Builders are literal; they like conversations full of details
and hard facts. But Explorers want to use these data to theorize creatively. So the Builder can come to believe the Explorer is intellectually careless, while the Explorer finds the Builder intellectually stodgy.

Moreover, the Builder is persistent—a trait the Explorer can regard as stubbornness. In turn, the Builder sometimes regards the uninhibited Explorer as fey and fickle, impulsively leaping to embrace
ways of doing things that have not been tested or authorized. The optimistic Explorer can also find the stoic Builder tedious and dour,
while the hardheaded Builder can regard the Explorer as unrealistic
and irresponsible.

Cautiousness can become an issue, too. I recall one Explorer
who headed for the ocean surf carrying his three-year-old son, in-
tending to introduce him to the thrills of big waves. When his Builder
wife saw where he was headed with their child, she leaped after them
with the rage of a mother bear.

Builders tend to be careful with their money. So the Builder will
probably want to save cash that the Explorer wants to spend. These
two won't always see eye to eye on planning, either. The Explorer
takes life as it comes, while the Builder needs control and enjoys planning way ahead. And these two types have an entirely different moral
compass. Builders tend to be strict about moral conduct, while Explorers often regard morality as flexible, based more on the circumstances than on a fixed moral code.

Last, this pair may disagree on how to spend their spare time.
Builders are eager to see their friends and do familiar things, while Ex-

plorers crave doing something new. So the Builder won't understand why the Explorer boycotts his or her parties, and the Explorer is likely to feel stifled by these engagements. For Builders, familiar pastimes are comforting; for Explorers, they are a curse.

Despite these differences, the Explorer and Builder both bring great strengths to the relationship. The Builder will ride out the crises and work like a beaver to stabilize the present and build the future. And the Explorer is flexible and positive; this fox is likely to shrug off the frustrations and hope for better days. Moreover, the Explorer can honestly appreciate the Builder's loyalty and strength of character, while the Builder may be continually drawn to the Explorer's curiosity, creativity and verve.

So if the Explorer and the Builder both continue to love and respect each other, even depend on their partner's traits to keep life balanced and fulfilling, their differences will be their strength.

Explorer-Director Match

This is a curious match. In my Mate Choice Study, these two types found each other mildly unattractive (see Appendix). Yet if they do fall in love, they will have some powerful things in common.

Importantly, Explorers and Directors can be highly compatible lovers—because dopamine and testosterone both stimulate the sex drive. Their attitudes about lust will probably differ, though: the Explorer is likely to take sex more casually, while the Director will imbue sex with all sorts of symbolic meanings. But both partners are likely to enjoy sex together regularly.

They are also likely to have interesting conversations. These two are often both intellectual. They both love playing with ideas. The Explorer, however, is likely to have far broader interests, while the Director will know a great deal more about fewer topics, focusing narrowly and deeply. Still, the Explorer will probably respect the Director's analytical, logical, rational mind. And the Director will admire the Explorer's curiosity, originality and breadth of knowledge.

I know a couple like this. John, a Director, is a biologist who knows a lot about genetics and brain systems. But to relax, he reads mysteries and watches sports on television. Linda, an Explorer, is an investment banker. In her spare time she reads widely: poetry, literature, the arts, the sciences, world affairs—just about everything except football interests her. John is deeply knowledgeable about a few highly complex and sophisticated academic subjects; Linda appears to know everything else.

Both these types can also be resourceful, even ingenious. And both are daring. But in the case of my friends John and Linda, John is very daring in his laboratory investigations—but nowhere else. Linda, on the other hand, travels the world to climb mountains, visit museums, taste new foods and mingle with the locals in alien, sometimes even warring countries.

Explorers and Directors tend to be self-sufficient; each will appreciate this trait in the other. And both are often very busy, professionally and otherwise. So neither partner is likely to be overly demanding of the other's time. Neither is likely to be possessive or unreasonably jealous, either. In fact, the Director's lack of possessiveness is comforting to the freedom-loving Explorer. And the Explorer's wanderlust or many local interests can give the Director time to focus on his or her deeper pursuits alone. Neither is inclined to follow rules or schedules, either—unless these strictures are useful. And both are flexible, open-minded, irreverent and unconventional. So neither is likely to force the other to follow traditions they don't respect.

These two types have very different ways of interacting with the larger world, however. As their name suggests, Directors tend to be direct, even blunt. They like to "get to the point" and say what they mean. Not Explorers. These men and women may appear direct at times, but they are chameleons who can change at a moment's notice to fit in with their environment. Nevertheless, the subtle Explorer is likely to find the Director's candor refreshing. And the Director can be charmed by the Explorer's free and easy way with people.

Where the Explorer can be flamboyant and hedonistic, the Director is emotionally contained and self-disciplined. But once again, each type can admire the other's style. The Director is likely to be drawn to the Explorer's energy and spontaneity; and the Explorer can enjoy loosening up the Director. Most important, both types like friends with whom they can have substantive and stimulating conversations. So neither will drag the other to superficial, tiresome social events.

Each of these two types can upset the other, though. Explorers are optimists, and they tend to respond impulsively to all sorts of things. Directors are far more skeptical, and they make up their minds with calculated, exacting reason. So the Explorer can find the Director's rational skepticism frustrating, even cynical and coldhearted. In turn, the Director can find the Explorer's impulsivity dangerous and chaotic. This difference can jeopardize their leisure activities, too. The Director admires self-restraint, while the Explorer tends to seek thrills.

Both types are competitive. But the Director is far more openly aggressive, something the Explorer—who doesn't wish to show his or her true colors—can't understand. The Director is also more goal-oriented and ambitious, while the Explorer is more relaxed and enjoys his or her many interests. So the Director is likely to think the Explorer is unfocused, even lazy, while the Explorer may come to regard the Director as tunnel-visioned. This difference in focus can also disrupt their intellectual exchanges. The Director can engage in conversations that sound to the Explorer like a tedious, academic lecture. By contrast, the Explorer can soar into flights of playful creativity that the Director finds silly and useless.

Yet this match can have great strengths, as I have seen in the marriage of my twin sister, Lorna, and her husband, Guy. Lorna is an Explorer, and Guy is a Director. Each brings many gifts to the partnership. She is a painter. And he is a genius at drawing up architectural and interior designs. When she has a gallery showing, it is Guy who measures walls, visualizes spatial relationships and hangs her work. He has also built beautiful homes for them over the years, designing elegant entryways, gardens, living spaces and patios. But Lorna, the Explorer, is the one who pilots them over the Alps in her balloon.

Explorer-Negotiator Match

The Explorers and Negotiators in my Mate Choice Study were not generally eager to meet one another (see Appendix). But female Negotiators were particularly allergic to male Explorers. Perhaps these men remind them of the "playboys" they dated in school.

Despite the typical lack of initial attraction between these two types, many do fall in love—due to timing, proximity and the subtleties of their love maps. And an Explorer and Negotiator can make a highly interesting match because of one remarkable aspect of their partnership: the Explorer is creative, and the Negotiator is imaginative. This is a wonderful combination for playing with ideas, something both types like to do.

As I've mentioned, I don't think creativity and imagination are exactly the same thing, although they are often complementary. Creativity often requires some imagination, and imagining is often creative. But you can create a piece of music or develop a theory of natural selection and never daydream about tomorrow or reflect on yesterday. And you can daydream endlessly without creating a poem, song, hypothesis or gadget. Nevertheless, the creative Explorer (particularly an Explorer of ideas) and the imaginative Negotiator are likely to get thrills from their mutual intellectual meanderings. And because the Explorer has many interests and the Negotiator takes the broad view, each will bring knowledge and insights to their conversations.

Both these types are also verbally skilled, although the Negotiator will be better at finding precisely the right words to communicate thoughts, while the Explorer will probably express opinions and ideas more spontaneously. Both types are also flexible, open-minded and agreeable. And both tend to be optimists. So both will revel in the breadth and speed of their conversations as they soar from one topic to another.

The Explorer will also admire the Negotiator's good-heartedness, win-win attitude and need for harmony and consensus. The Negotiator, meanwhile, is almost certain to appreciate the Explorer's sexual appetite. Negotiators believe that sex is important to a partnership; but they can get caught up in so many other activities and concerns that they don't initiate it. In the Explorer, they have a sensual and ever-ready physical companion. Moreover, the Negotiator may add meaning and emotional depth to the Explorer's playful times in bed.

Negotiators are romantic. They like to create loving rituals to keep the passion sparkling—such as a special good-bye morning kiss, surprise greeting card or getaway weekend. Explorers will enjoy these gestures; to them, romance requires an element of play. Negotiators tend to show their emotions, too. So although Explorers are often secretive about their deepest thoughts, they can be intrigued by the emotional transparency of their Negotiator mates, even feel invigorated when swept into these storms of feeling. The Explorer is probably also amazed at how trusting and patient the Negotiator can be; Explorers are far more fickle.

But after some of the romantic fervor begins to blend with the routines of a daily partnership, these two types can develop problems.

The Negotiator seeks to understand his or her own identity, as well as delve into the psychological underpinnings of the relationship. The Explorer can become cynical about this need for introspection, and impatient with the Negotiator's desire to bestow meaning on every tiny interaction between them. Meanwhile, the Negotiator may begin to yearn for more intimate discussions about feelings and motives, and may even come to believe that his or her Explorer partner— who looks out, not in—is shallow.

A related problem can stem from their different needs for autonomy. Both like their space. Both are independent-minded. Both respect the autonomy of the other. But the Negotiator wants to have an authentic and deeply intimate connection with a mate in a way that the Explorer may never understand. And the Explorer needs a genuine kind of freedom that the Negotiator can find mystifying. So the Negotiator may feel more and more emotionally abandoned while the Explorer feels stifled by this partner's need for closeness.

Negotiators are also particularly skilled socially, generally far more so than their Explorer mates, as well as more empathetic and intuitive. Hence Negotiators are more likely to take into account their mate's needs as they make decisions, sometimes going to great lengths to please. But Explorers don't always appreciate this nurturing. Carefree and tough-minded, they can misconstrue this compassion, regarding their partner as indecisive, even spineless. Meanwhile, Negotiators may wonder why a mate is insensitive, indeed incapable of reading between the lines to understand the nuances of the partnership. And unlike Explorers, who tend to forget quarrels quickly, Negotiators remember. These men and women can nurse a grudge, too.

Negotiators are idealistic. They care not only for their partner but also for everybody else. So when making decisions, they are likely to make an unrealistic choice to avoid hurting someone's feelings. And although Explorers are equally egalitarian, equally unimpressed with rigid or unfair rules and regulations, and equally disdainful of authority figures who take advantage of others, they don't share the Negotiator's global empathy. Sometimes they can find their mate overly sensitive, even weak. The Negotiator, meanwhile, may come to find the Explorer hard-hearted and self-centered.

As with other matches, however, both the Explorer and Negotiator have traits that can ease the partnership through troubled times. Perhaps most important, the Negotiator has a deep need for harmony, while the Explorer is tolerant. As a result, both are inclined to overlook the hardships and persevere. Both are also flexible. Both crave romance. And both feel vitally alive while enjoying their creative and imaginative conversations and adventures. Moreover, the decisive Explorer can aid the indecisive Negotiator, and the compassionate Negotiator can love the more fickle Explorer unconditionally.

So if each partner retains their positive illusions about the other, they can maintain their pas de deux for years.

Builder-Builder Match

Builders are far more attracted to other Builders than to any other type, as was clear in my Mate Choice Study (see the appendix). They are also more likely to marry one another, more likely to say they are happy in their marriage and less likely to divorce. This is a highly compatible—and natural—match.

Builders tend to be calm, relaxed and socially skilled. They are also literal and precise. So two Builders can enjoy casual chitchat, sharing facts about sports, people, health, politics and business. Builders are managers, too. Both like bringing people together, building community ties and maintaining their social networks. And because Builders are self-effacing, affable, civic-minded and often popular, a pair of Builders will enjoy working together to create a wide and stable social network. They will be effective, too—and share a lot of close friends.

These men and women are not impulsive with their money, their actions or their feelings. Security is important to them. Builders have a clear memory of yesterday's mistakes, so they prepare for the future. As a result, two Builders can respect each other's need to save money and spend it carefully.

I know one couple like this, Laurie and Ralph. Both have worked hard; both have saved their money. Now they're older, and both are eager to travel and enjoy some of life's luxuries. But they are having trouble parting with their cash. Recently Laurie told me of a touching moment between them that occurred one morning as she left their house. At the door, Ralph hugged her and whispered, "Spend a little money, sweetheart." They laughed. She didn't.

Builders are traditional. Duty, loyalty and service to the community are the Builder's strong suits. And two Builders will respect each
other for their shared values. Moreover, both will be devoted to family
activities and will work to sustain domestic stability at almost any personal cost. Both are also patient, persistent and industrious, as well as
willing to do thankless jobs around the house. And both like to make
plans and stick to schedules. So both will overlook their differences to
follow the rules and make the marriage last. Fortunately, Builders
don't get bored easily either, which helps them keep their vows.

But two Builders will encounter various problems as their relationship matures. First of all, both are moralistic. Both believe there
is a "right" thing to do and a "right" way of doing it. Both can also be
critical. So two Builders are capable of having an argument over trivial matters, such as how to fold the laundry or wash the dishes.

They can also lock horns over important issues, as is the case
with Sam and Charlotte, a happily married Builder couple and friends
of mine who took my Personality Type Test. Sam is a BUILDER/
Director; Charlotte is a BUILDER/Negotiator. They met at a dance in
a small town in Virginia when he was sixteen and she was fifteen. Both
were with friends. As Sam said, "We started talking, and we've been together ever since." They have now been married twenty-seven years.

They have three issues in their marriage. One is how to raise
their children. As Charlotte says, "I am more of the caretaker, while
Sam is into 'tough love.' If one of our sons wants to go out with
friends, for example, Sam makes him do something to earn his free
time. I don't think that's right. I think the kids should go and have fun,
and then when they come home they can help out if I need them."

Arguing over child rearing is common among spouses. But like
many Builders, Sam and Charlotte refuse to compromise their personal standards; both are stubborn. In this situation Charlotte and
Sam are also expressing the secondary aspects of their personalities.
Sam is being tough-minded, a trait of the Director, while Charlotte is
more flexible and contextual, traits of the Negotiator.

Another issue between Sam and Charlotte is intimacy. Both

agree that they don't spend enough "quality time" together. As Charlotte says, "When Sam is home I still do my daily routine, and he does
his own thing. It's almost like we're on different schedules." Sam
agrees, saying, "Charlotte and I have our own routines, and we don't
include each other enough." Builders feel that routines, schedules and
traditions keep life orderly and relaxing. In Sam and Charlotte's case,
their mutual need to follow their personal schedules is endangering
their closeness.

Builders are also the least sexual of the four types. So I wasn't
surprised to learn that lack of sexual intimacy was Sam and Charlotte's third problem. Sex—and the closeness that it can generate—
had also taken a backseat to their schedules.

In spite of these issues, both Sam and Charlotte report that they
are still very much in love. As Charlotte puts it, "Even though Sam
and I have a couple of kinks to work out, I am still so in love with him.
Sam treats me well, he's very smart and talented and I trust him one
hundred percent. I think he's my soul mate."

Notice Charlotte's use of the words *soul mate,* an essential need
of the Negotiator. But in Sam, Charlotte also has a man who is likely
to be loyal to her all her life. I suspect she will remain happy with him,
too. In my 2007 study for *Oprah* magazine, I found that couples composed of two Builders (most of them married more than fifteen
years) were particularly happy with each other.

Builder-Director Match

Love at first sight is not likely to occur too often between a Builder
and Director—because the men and women of both types avoid
meeting each other (see the appendix). On Chemistry.com, they look
at one another's pictures, read one another's words and regularly re-
ject these individuals. Of course, some Builders and Directors do fall
in love and marry. But they are probably attracted for reasons other
than biological compatibility. Perhaps the timing is right, or they
work or live nearby, or each provides much of what the other needs.

Both may also cling to their positive illusions about their partner—
because these types are vastly different.

Builders tend to be socially skilled, while Directors are straightforward. So if a Builder and Director can respect each other's social
style, they can make a strong team. The Builder can even balance out
the less socially adept Director, and the Director can add a forthright
note to their conversations. But if the Builder becomes irritated by
the Director's "rude manner," and the Director becomes impatient with
the Builder's "superficial chat," these partners will have trouble at the
breakfast table.

Builders tend to be social, too. To them, community and business networks are important. They want to fit in, to belong and to
nurse their connections with others. Directors, on the other hand, are
rarely interested in cultivating social networks. To these men and
women, social mingling is a waste of time unless the people are very
interesting or very useful to their career. So rather than building extensive social networks, Directors choose to make a few close friends.

Fortunately, these dissimilar social interests aren't always a disas-
ter. The networking Builder can broaden the Director's social circle,
while the Director can add a handful of dedicated friends to their social mix. Still, the Builder is likely to initiate parties and outings that the
Director can't abide, then feel abandoned when the Director shows up
late or not at all. The Builder may even feel that the Director is trying
to destroy the social networks that he or she is working so hard to
maintain, and pine for a partner who will engage more enthusiastically
in his or her social world. Meanwhile the Director may feel pressured
to engage in what he or she regards as meaningless social pleasantries.

Another cause of discord can arise from their very different interests. While the Builder tends to have broad social interests, the Director often has deep and narrow intellectual pursuits. In a good
relationship, the Director's various hobbies and work interests can
bring intellectual vigor to the Builder's more relaxed social world. In a
sour partnership, the Director will pursue these interests alone, and
the Builder will feel excluded.

Even their conversations tend to be entirely different. Builders like to stick to facts and details, while Directors like to analyze these data, then use this information to build theories. Moreover, the Builder finds it relaxing to exchange unrelated facts and trivia about the weather, sports or people, which can bore the Director. The Director, meanwhile, is eager to have discussions on technical or esoteric subjects that are of no interest to the Builder.

These two can butt heads over many other issues as well. Builders respect authority and pride themselves on following the social rules. Not Directors—these men and women scorn authority. They refuse to obey conventions of any kind, unless doing so suits their needs.

These two types also have quite different ideas about morality. Builders have a fixed moral code. And they can dig in their heels, unwilling to change their perspective, their habits or their values. For Directors, the "right" thing depends on the circumstances; moreover, they are determined to do what is most expedient.

Finally, while the Director is highly competitive, the Builder is group-oriented. So when the Director launches a debate at the dinner table, or even an outright verbal attack, the Builder uses all of his or her social skills to duck what seems like intellectual warfare.

Despite these major differences in temperament, the Builder and Director can work well together in the home. Both are eager to create and maintain a stable family life. Both are orderly, meticulous and efficient. Both are usually emotionally contained. Both are realistic. And each will take on complementary household tasks. The Builder is likely to do more of the daily work necessary to keep things running smoothly, like making plans, following schedules and buying the necessities to create a comfortable home. The Director is likely to focus on stabilizing the family income instead.

The Builder and Director will also be good at pooling their talents to make family decisions. The Director is bolder than the Builder, but exacting; he or she will make decisions with a thorough and confident style that the more cautious Builder can admire. And both types tend to be loyal. In fact, both will work to maintain a less than perfect partnership—although the Builder may be more motivated to sustain the relationship to maintain his or her respect within their community, while the Director may wish to avoid the personal and financial disruptions of divorce.

These types also know how to have fun together. The Director is likely to be more sexual, bringing vitality to the bedroom. Moreover, the Director tends to be more creative and daring—traits that can excite the more traditional Builder. And the Builder's relaxed attitude and social charm can calm the Director.

These types have another powerful complementary trait. Directors tend to be technically skilled, even inventive. Many of humanity's greatest inventions have sprung from the minds of Directors. Builders, on the other hand, have the managerial abilities to develop these ideas. I suspect that many successful companies are launched by Directors, then run by savvy Builders. So if this pair can combine their inventive and managerial talents to run their home, rear their children and enjoy each other's company, the Builder and Director can march forward like two feet—and achieve great happiness together.

Builder-Negotiator Match

Here's another pair that isn't naturally drawn to each other (see Appendix). Still, many Builders and Negotiators do fall in love and marry. And if they can use their differences to their advantage, they can make a successful match.

These two types, for example, have very different attitudes about themselves. Builders have a firm view of the world and their place in it; deep introspection doesn't interest them. Inward-looking Negotiators search for their own identity instead; they thoroughly question and analyze their values, worth and objectives. So the Negotiator is likely to be attracted to the Builder's dependability and self-assurance, while the Builder, a guardian by nature, can enjoy being a beacon of stability and self-acceptance for the Negotiator.

Negotiators see many sides to every issue, many avenues to any goal. Hence they can get lost in the big picture, procrastinating as they imagine and reflect. Builders, on the other hand, think concretely, carefully and cautiously; then they act decisively. Indeed, they construct meticulous schedules and deadlines to get things done in a precise and thorough manner. So the Builder provides a secure world in which the Negotiator can more comfortably make important family decisions and navigate internal storms.

The Negotiator, on the other hand, is enthusiastic, optimistic and lively. These men and women are also more curious than Builders and have more interests. They like to pursue every aspect of their interests, too. So the Negotiator can energize the calmer, less curious Builder—bringing fresh ideas to the partnership. As a result, the Builder is likely to find the Negotiator invigorating, if somewhat naive.

Such a couple is Kent and Nora, whom I met recently while staying in their bed-and-breakfast in the Outer Hebrides. Nora, the Negotiator, owns the inn; Kent, the Builder, works on an oil rig in the North Sea in two-week shifts. During my visit, he was off the rig, and graciously took me around to see the local sites. At one point he said of Nora, chuckling with pride, "She's interested in *every*thing." He takes her annually to New York to shop and see the sights, although he is far less curious about the city than she. I suspect he needs her optimism, energy, curiosity and creativity to feel alive. She, I'm sure, appreciates his steadiness and consideration.

Both Builders and Negotiators are eager to build strong, stable relationships. Negotiators are far more emotionally expressive, however. They will tell their Builder partner how they feel, as well as search to understand their mate's deepest needs. Although the Builder may not appreciate the depth of the connection the Negotiator strives to make (no type does, except another Negotiator), he or she is likely to be drawn to the Negotiator's emotional depth. Meanwhile, the Negotiator can enjoy establishing a heartfelt connection to a partner they feel will be loyal to their eternal bond.

Builders and Negotiators share a deep interest in home and family.

Both are nurturing, although Negotiators will probably be more sensitive to their partner's needs. Both can be orderly, although the Builder is likely to excel in this respect. And both will be attentive to their children. The Negotiator is likely to be far more lenient, however, while the Builder will be stricter about rules and schedules. Both will enjoy planning for the future, too.

The Builder and the Negotiator also have a similar interest in "belonging." And both are socially skilled, so this pair can work easily together to develop strong business and community networks. Moreover, both are eager to help others, although the Negotiator is likely to be more compassionate toward individuals while the Builder is more concerned with supporting the community. Still, each type will value the other's moral seriousness. These two types also tend to feel strongly about God. The Negotiator is likely to take a personal, unconventional spiritual path, however, while the Builder may be more drawn to organized religion.

Yet like other matches, this pair will have its share of rough times. The emotionally contained Builder can become uncomfortable with the intense passion and sweeping emotional gestures of the Negotiator. The Builder can also be mystified by the Negotiator's need to explore his or her identity, the larger meanings of life and esoteric philosophies, regarding these pursuits as frivolous. Moreover, the Builder can be critical of the Negotiator's imaginative ideas and new ways of doing ordinary things, innovations the Builder can come to see as threats to cherished traditions and routines.

I know one couple who used to have a fight every time they cooked a steak together, even though both liked their meat very rare. The smallest change in the routine would set them off, the Builder wanting things done the "normal" way, the Negotiator insisting on trying a new culinary technique.

For their part, Negotiators can become disappointed because their Builder partner has proven to be less romantic than expected. Negotiators can yearn for more depth and meaning in their relationship, too, things the practical Builder has difficulty fathoming, let alone providing. Negotiators may even find the Builder superficial and try to remodel this "down-to-earth" mate. For a while, the Builder may try to satisfy the Negotiator's thirst for more in-depth conversations about "the relationship." Eventually, however, the Builder is likely to feel unappreciated and irritated with this partner's seemingly foolish self-absorption and need to analyze the psychological underbelly of their lives together.

Builders and Negotiators have other basic differences that can disappoint. Builders like to stick to the facts and focus on concrete reality. Negotiators, by contrast, are intuitive, and they love to speculate. So although the Negotiator may be initially attracted to the Builder's realistic approach, eventually this "facts only" attitude can get on the Negotiator's nerves.

Negotiators are not particularly dedicated to traditional values, either. These men and women do not respect authority without questioning it first, and they do not believe people must always behave according to established patterns of "proper" conduct. Builders, however, believe that long established customs need to be respected and preserved, and that authority should not be questioned unless there are powerful reasons to do so. This single difference in temperament can lead to real conflict.

These two types will sometimes drive each other crazy with the way each reasons, too. Builders stick to the data and the facts when making decisions; they are not swayed by mere opinions. And they don't want to be deflected by ancillary considerations; they like to work in a straightforward path directly toward their goal. Not Negotiators. They fly off on tangents of imagination and weave long webs of possible repercussions before they make any decision. So while the Builder can come to regard the Negotiator as featherbrained, the Negotiator can come to believe the Builder is unimaginative and boring.

In spite of their differences, both the Builder and Negotiator are nest makers. And when they work together, the Pillar of Society and the Philosopher King can make a loving and effective team.

Director-Director Match

Recently I had a business appointment with a corporate lawyer in a breakfast room of a Boston hotel. Within seconds of meeting him I knew he was a Director. He was tall and had the rugged, handsome face built by testosterone. He was commanding in his black pin-striped suit and red tie. He was also direct, focused, tough-minded and aggressive. Indeed, I thought he just might order a bowl of rocks for breakfast.

However, he knew about my personality research, and after some preliminary conversation, he told me that he had married another Director, someone like himself. They were both highly successful lawyers, he said; and both were very analytical, exacting and decisive. Summing up his relationship with her, he said, "I finally found a woman I understand. She thrills me."

Directors are not particularly drawn to each other at first sight; but they have no initial dislike, either (see the appendix). As they look at the photos and read the words of other Directors on Chemistry .com, the women respond neutrally to the men, and the men are just slightly positive about the women. Nevertheless, when two Directors fall in love, they can be thrilling company for each other.

Directors are often very competent. And they are drawn to anyone who is equally focused and shares their concerns. So two Directors almost instantly begin to talk in great detail about their shared interests. Moreover, both are analytical, logical, skeptical and critical. Both like to stick to the data, rather than expose their needs or feelings. Both like to theorize. Both like to debate. Both say what they mean. And both appreciate the other's direct, informed and exacting approach.

These men and women are the most emotionally contained of the four types. They like to solve their problems without letting their emotions get in the way. So other types can find Directors aloof, detached, even cold, but two Directors understand each other. And they know that beneath a sometimes frosty exterior boils a human being with real feelings and real needs. Moreover, because neither tends to engage in false social gestures, both know that when they express compassion, apologize, ask for help or say "I love you," they mean it.

Directors are also decisive. But as you know, they do not make their decisions impulsively. They analyze the situation logically and calculate the outcomes thoroughly; then they act. So although a pair of Directors may appear cautious, they are actually logical and meticulous. And each partner can admire this strategy in the other.

Both also scorn social conventions. Directors are not traditional. Neither will drag the other to what both regard as frivolous social occasions. Instead, each will introduce the other to a few close friends. And both will enjoy sharing intimate evenings with like-minded people eager to have intense conversations on interesting topics. Both will also enjoy the impromptu late night movie, quick trip to somewhere for a two-day vacation, or an early morning walk together—as long as these spontaneous times don't interfere with either's work or hobbies.

Directors are ambitious and competitive; they often want to get to the top of their field, and they have the energy to work at it. So they are dedicated to their careers or other pursuits, and they understand this commitment in each other. Both are highly independent, too. And neither is overly possessive. Each can therefore give the other the time and privacy they need to get ahead. In fact, these men and women can tolerate, even enjoy, long periods of time alone, working. Neither needs to renew the partnership with constant verbal expressions of intimacy.

What's more, sparks are likely to fly in the bedroom. Directors tend to have a high sex drive. And although a pair of Directors are not likely to be emotionally expressive at the breakfast table, or even during a candlelight dinner, both can open up while making love. In fact, Directors regard sex as deeply intimate and a precious gift to be given and received. And each appreciates this attitude in the other.

But there are potential problems in this partnership, as in every other match. To begin with, Directors tend to be workaholics. Even when working at their hobbies they can become overly absorbed. So after the initial, passionate romance has been tempered with feelings of secure attachment, Directors often don't make sufficient time for each other, a bad habit that has been directly associated with high levels of testosterone.

Which reminds me of Ted and Jackie. Both work too hard, he as an academic psychologist with many students and projects, she at her retail business. Both had slipped into the habit of barely seeing each other. Finally they decided to view their marriage differently, regarding it as a threesome: Ted, Jackie and "the Relationship." Now every Saturday they go off and do something exciting together—to please and satisfy "the Relationship." This fictitious third party helps to keep the marriage strong.

Just as toxic as ignoring each other, two Directors can insult each other. These partners enjoy matching wits. So they can compete when they play sports or card games, or even when completing crossword puzzles. Both are also skeptical and exacting. And both can be aggressive, even rude: two Directors sometimes question each other's facts or logic when they talk, or hurl criticisms that make each other angry. They will return the attack, too. Neither wishes to be intimidated. Both want to win.

As a result, two Directors can fall into a pattern of rarely spending time together, misunderstanding each other when they do and being critical when they feel disrespected—which only makes their misunderstandings worse.

But when two Directors appreciate each other's thirst for knowledge and achievement, spend enough time together and argue constructively, both can be devoted partners. Directors can be very loyal. Besides, Directors don't like to fail at anything, including their primary relationship. Many don't want to face the nuisance of divorce or dating again, either. So two Directors are likely to ride the bumps to keep the partnership intact. They may even find a consistent and long-lasting thrill in the relationship—particularly if each regards the other as a mind mate.

Director-Negotiator Match

Although Negotiators and Directors have extremely different temperaments, these types are often drawn to each other at first sight, as my Mate Choice Study showed (see the appendix). This is a natural complementary match.

To begin with, their thinking meshes. Both types of people dislike wasting time on irrelevant or superficial conversations. Both like to discuss abstract concepts and generate theories. And both enjoy conversations about philosophy, literature, the arts, the sciences or politics. Yet with their web thinking, Negotiators see the big picture, while Directors focus intently on a smaller piece of the puzzle. So each type can impress, even enchant, the other with their knowledge and insights.

They also make decisions well together—each adds skills the other lacks. Negotiators have a complex inner life full of possibilities and ethical dilemmas; they can think of many facets to a problem and many ways to solve it. As a result, they are inclined to ruminate without acting. Directors are decisive yet likely to skip over what they regard as ancillary data in their need to achieve their goal: the solution of the problem. The Negotiator sees all the angles, and the Director can make the final decision. It's a great team for reaching an effective solution to most problems.

They also have complementary social skills. Directors are the least compassionate of the four types; Negotiators have the most empathy. Directors are poor at verbalizing their thoughts and feelings; Negotiators excel at all sorts of linguistic skills, particularly at finding the right word rapidly. Directors are emotionally contained; Negotiators are emotionally expressive. And if Directors are too blunt, forthright and competitive, Negotiators are too eager to please. So in social situations, the Negotiator can smooth over the Director's inappropriate comments, while the Director can contribute some refreshing candor to the conversation.

Each of these personality types can admire the dissimilar qualities of the other type, too. Negotiators pick up every nuance of every social interaction; every tilt of the head or lilt in the voice speaks volumes. Hence, for them, the Director's bluntness can be enjoyably easy to decipher. Negotiators also like the Director's ability to focus on one thing at a time, capacity to concentrate on their goals and gift of knowing his or her own mind. And Negotiators appreciate the Director's ambition and willingness to sacrifice to get ahead.

It's mutual admiration, too. Directors are drawn to the Negotiator's enthusiasm, warmth and insight into people. Directors are comforted by the Negotiator's compassion and heartfelt approval, even if they don't admit it. And even though Directors are the least able of the four types to change their minds, they can appreciate the Negotiator's mental flexibility.

Moreover, Directors can become so focused on work that they sometimes miss life's basic pleasures—an impromptu Sunday picnic or their child's school play—unless their mate insists they participate. Negotiators know how to influence Directors; they are diplomatic yet firm. This trait is important to Directors, who want a mate who has the strength of character to stand up to their formidable personality. In the pacific, charming but determined Negotiator, they find their match.

Both types also value personal autonomy. But the Director is far better at expressing his or her need for independence, whereas the Negotiator is more eager to please. So the Negotiator can admire the Director's ability to act freely, while the Director can appreciate the Negotiator's agreeableness. And each can enjoy the trust and freedom the other provides.

The Negotiator and Director mesh in many respects, but perhaps most important is the way they court and sustain their love. Nothing holds either back from pursuing what they regard as a worthwhile relationship, then deepening the commitment. Both types are capable of reaching extreme intimacy with a beloved, too, although the Negotiator is more verbal about it.

Problems can emerge, of course. Directors are not emotionally expressive; instead they strive for efficiency and logic, traits that can disappoint the expressive, often sentimental Negotiator. Directors don't like redundancy either; they are not likely to say "I love you" on a regular basis, or to respond to many of the other tender rituals the Negotiator is likely to create to feel connected. So trouble will surely start if the Negotiator begins to construe the Director's aloofness and preoccupation with work as thoughtlessness or indifference. And as the Negotiator begins to feel unnoticed and unloved, the Director can feel misunderstood.

Early in their marriage, neuroscientist and Nobel laureate Eric Kandel and his wife were a good example of this particular stress and misunderstanding between two very different personalities. In his autobiography—which suggests that he is a Director wedded to a Negotiator—Kandel writes of a pivotal moment: "Shortly after our son, Paul, was born, in March 1961, Denise and I had a serious crisis, by far the most serious of our life together. I thought we had an unusually harmonious relationship . . . But one Sunday afternoon she showed up as I was working in the lab and simply exploded on me. Carrying Paul in her arms, she screamed, 'You can't go on like this! You are only thinking of yourself and your work! You are just ignoring the two of us!' I was startled and deeply hurt. I was so transfixed by my science . . . that it never dawned on me that I was . . . withdrawing my love from them . . . In response, I decided to spend more time at home."

Kandel overcame this classic conflict between a Director and a Negotiator. But this is not the only major issue that can jeopardize this partnership. Directors admire self-control. So if the Negotiator turns into a drama queen (or king), the Director is likely to retreat emotionally, deflecting further discussions with hostile silence. The Director can also become impatient with the Negotiator's intuitive side, regarding it as airy nonsense. Meanwhile, the Negotiator may begin to find the Director's insistence on logic irritating, particularly if the Director's proclivity to do "what works" tramples on someone's feelings.

These personalities also differ in how they look at life in general. The Negotiator is far more idealistic. As a result, the Director can become annoyed by the Negotiator's far-flung humanitarian concerns, while the Negotiator can begin to regard the Director's more technological approach to fixing the world's problems as narrow-minded and unfeeling.

This match is often regarded as the classic example of how "opposites attract." You probably know plenty of them: the hard-driving, aggressive, blunt man married to the socially skilled woman who runs his home and supports his harried business life. But female Directors, such as Hillary Clinton, also marry their complementary type. To balance her direct, decisive, focused personality, she married a talkative man with a warm, open, synthesizing mind.

When a Director and Negotiator put their heads and hearts together, they can make an exciting and formidable match.

Negotiator-Negotiator Match

Negotiators are true idealists, the least realistic of all four types regarding romance, love and marriage. They seek the perfect partner and eternal love. And they refuse to compromise. Some Negotiators would rather live alone for years than settle for anything less than a deeply meaningful relationship. Nevertheless, Negotiators are not generally attracted to each other at first meeting. My Mate Choice Study showed that Negotiator women have only a very mild initial attraction to Negotiator men, while Negotiator men tend to avoid women of their type (see the appendix). When two Negotiators fall in love, however, they are likely to revel in a dream come true.

Negotiators want to talk to their partner about everything: how their partner feels; what their partner thinks; their mate's goals, hopes and disappointments; their partner's childhood; and particularly the intricacies of their life together. Negotiators reach out. But they also reach in, analyzing their own motives, thoughts and feelings. So two

Negotiators can talk endlessly about each other and themselves. Both want to share and connect.

With their superb communication skills, two Negotiators can travel into each other's psyches easily, too. These people are intuitive, picking up a host of subtle messages from the other's slightest frown, shift of body weight or change in tone of voice. Both can also express their feelings vividly and accurately, with just the right words. And each tends to be highly sensitive to the other's feelings and needs. So two Negotiators can shower each other with attention, understanding and forgiveness.

Two Negotiators have many things to talk about, too. Both see the big picture. Both accumulate information and assemble it into complex webs of thought. Both like to generalize, synthesize, play with ideas and digress into side conversations. And both are highly imaginative, creating all sorts of amusing or intriguing scenarios from just a few facts. Not surprisingly, this pair often engages in ingenious and inspired dialogues—swinging easily from facts to broad theories about how the world works and why people do the things they do.

Neither is a social butterfly, either. Nor are they impressed by social rules, standards of proper conduct or long-held traditions. Like all but Builders, Negotiators are not conventional. Instead, they strongly believe that each ethical, social and political situation deserves special consideration, requiring its own moral approach. What is most important to these men and women is their own conduct: they must be genuine and honest. Interactions with friends must be personal and authentic, too. So both are likely to introduce their partner to friends whom the other will like and respect.

Negotiators are romantic. They go to great lengths to sustain feelings of passion for their sweetheart. For example, I recently met a woman, Pamela, who told me that she regularly takes a long subway ride in New York to buy bottles of a special salad dressing for her husband-to-be. She whispered to me, "He likes it so much he can even drink it out of the bottle while he watches TV." Negotiators call each other from anywhere in the world to stay in touch. They bring home thoughtful gifts or compose poems or songs to serenade the other. Both want to bring beauty, passion, intimacy and affection to the partnership.

Negotiators also share beliefs. Both tend to be spiritual, if not formally religious. Both are also likely to have a deep concern for the local or world community. And both tend to devote time or money to their personal social causes—be it saving polar bears, educating the underprivileged or supporting cultural, educational or political institutions. And each respects this idealism, altruism and activism in the other.

Despite what may sound like an ideal match, however, this partnership can encounter some stunning pitfalls. To begin with, Negotiators sometimes give each other too little privacy—and get on each other's nerves. A Negotiator friend of mine, Laura, smothered a relationship in exactly this fashion.

Laura and her new Negotiator man, Jeff, fell into each other's arms one evening at a party and soon became inseparable. They rarely went out; instead they lay on her couch endlessly baring their souls to each other. Soon every nuance of their interactions became swollen with subtle meanings. "Sometimes a cigar is just a cigar," Freud famously said. Not to these two Negotiators. Moreover, they thought they could surmount any problem with honest discussion. So when Jeff began to want to make lengthy journeys to take part in esoteric musical events and Laura didn't want to accompany him, they mutually decided that he should go with a different partner, his former girlfriend. But too much sharing and compromise soon choked their feelings for each other. The relationship imploded from its own psychological weight.

Negotiators also have trouble making up their minds; they become swamped in the endless possibilities. And when they do make decisions, they tend to make them from the heart, because they are loath to hurt anybody's feelings. Hence this pair can be slow to solve a problem and too soft-hearted to make the most logical decision when they do. Both are also mechanically challenged. Two Negotiators will have to hire others to fix the lawn mower or set up a new computer. Neither can handle any kind of personal confrontation, either; and neither is direct. So when relationship troubles loom, each is likely to lash out by gossiping behind the other's back.

Most Negotiators are very slow to leave a partnership, however. They ruminate instead of acting. Negotiators feel such compassion for their mate, as well as such unease with conflict, that they go to extremes to address the problems in their relationship instead. But they won't procrastinate forever. If two Negotiators can't find genuine intimacy with each other, eventually their disillusionment overcomes their anxiety and they turn away to search again for eternal love.

Still, many Negotiator couples find—and keep—what they have sought all their lives: a soul mate.

Nature's Plans

Poet W. H. Auden once said, "Any marriage, happy or unhappy, is infinitely more interesting than any romance, however passionate." I agree. A marriage is like an oriental rug, a richly woven tapestry of common goals, shared interests, respect, humor, patience, compromise and dogged determination. No two partnerships are exactly alike. And every combination of personality types has its merits.

Nevertheless, certain types of matches have some natural assets. An Explorer will find a play mate in another Explorer. A Builder will find a help mate in another Builder. A Director will have a mind mate in the Negotiator, and a Negotiator can find a soul mate in the Director. These partnerships have some innate strengths. Other matches are more challenging, the most problematic being those between the Builder and any other type but another Builder. Builders have strict habits and distinct values that the other three types don't share. Two Directors and two Negotiators are likely to have a different problem: these partners are often so alike that their weaknesses can overcome their strengths.

Nevertheless, many forces other than biological temperament help to sustain a romantic relationship. If your beloved comes from your background, is the right age, has your level of intelligence and education, is humorous and socially adept, shares your religious and social values, has interests that interest you, can satisfy your needs, offers an appealing lifestyle and fits within your concept of an ideal mate—your love map—you may keep the sparks flying for years.

Indeed, we often *become* the person our partner believes we are. He might think she's humorous, for example, so she drops her inhibitions and becomes funny. She thinks he's generous, so he becomes munificent. Psychologists believe that among happily married couples, each partner begins to assume the personality their mate envisions—and this idealization begins to shape the relationship in positive ways.

Perhaps this was the case with the Mayan sun king, Hasaw Chan K'awil. He continued to believe his wife was the finest woman in the empire. And she became it. But I suspect their natural biological temperaments were also in harmony.

Come to the edge, A said.
I'm afraid, B said.
Come to the edge, A said.
I'm afraid, B said.
Come to the edge, A said.
And B came to the edge.
Jump, A said.
And B jumped.
And B flew.

—ANONYMOUS

PUTTING CHEMISTRY
TO WORK:

Advice from Nature

People ask me whether knowing what I know about romantic love has spoiled it for me. Hardly. You can know every single ingredient in a piece of chocolate cake and still get that rush of joy when you eat it. So it is with love. Romantic passion is always exhilarating. The delicious pleasure of being with someone you think is special and unique; your feelings of ecstasy, energy and optimism; your intense focus on everything your beloved says and does; your possessiveness and maniacal determination to win this bewitching creature: all swamp the rational mind as you become electrified by craving and obsession. Someone wonderful is camping in your head.

"Lovers and madmen have such seething brains," Shakespeare wrote. He was right; we now know what romantic love is, at least partially. My colleagues and I have put forty-nine men and women who were madly in love into a brain scanner and recorded the activity in their minds as they looked at a photograph of their beloved. Seventeen were happily in love; fifteen had just been rejected in love; and seventeen reported that they were still in love after an average of twenty-one years of marriage.

Several brain regions and circuits "lit up." Prominent were parts of the brain's "reward system," the complex network that produces elation, yearning, motivation and addiction. In fact, one of the brain regions that becomes active when you first fall in love is also triggered when you feel the rush of cocaine.

Wild is love.

But romantic love is much more than a cocaine high. Around the world people sing for love, write about love, pine for love, live for love, kill for love and die for love. Romantic passion is one of the most powerful circuits in the brain.

So I suspect that all four personality types are equally helpless when they love—as are young and old, gays and straights. In fact, when I queried 837 men and women in America and Japan about love, the gays and lesbians in the group showed just as much romantic passion as did the heterosexuals. And I know children as young as five and men and women in their eighties who are besotted. Everybody loves.

How do we ignite this madness in ourselves or in another? How do we keep this passion burning? I am not a therapist. But I have come to understand some of nature's ways. So here are some tips for each personality type and their lovers.

The Explorer

"Exhaust the limits of the possible," advised the ancient Greek poet Pindar. Explorers do. Explorers are drawn to people spontaneously. In fact, Explorers can be so impulsive that they get romantically involved too quickly. Even when Explorers seek a long-term relationship, they are likely to do it hastily.

So if you are an Explorer, *go slow.* Don't get into a relationship too fast.

Because Explorers are spontaneously generous, they tend to give presents too soon, too. I know a man who recently gave a woman two dozen long-stemmed red roses just for having an exciting luncheon conversation with him. She thought he had fallen in love with her, and she was bitterly hurt when she discovered he was madly in love with someone else. Flowers; a book; tickets you can't use: don't give even little presents to someone you hardly know unless you are interested in beginning a romance or make your other intentions clear.

Resist the grand gesture of paying the dinner bill or throwing a lavish party for a new lover and all of his or her friends. It may lead to unwanted expectations.

And don't make promises you can't keep. The other three types are likely to take you more seriously than you intended.

Explorers crave autonomy, and they can misinterpret another's genuine concern for them, regarding it as controlling. Relax. Some people feel better about *themselves* when they show concern for *you.* So if someone wants to help you, accept the offer.

Don't be too wary of a would-be partner who appears more cautious and traditional than you. Underneath your date's restrained exterior may be a delightful individual who wants to go adventuring and seeks an Explorer to point the way.

Don't dismiss a calm, orderly, trustworthy person as unexciting either. He or she may be just what you need to make a happy home.

Avoid talking to your date about other women or men with

whom you have been romantically involved—even if you have a hilarious tale to tell or wish to impress your date with some spectacular moment in your past. Almost no one is comfortable listening to all the details of a former relationship.

Keep in mind that there's no way to size up a person's sexual skills or interest by looking at them. Don't leap to conclusions about the sexual prowess of a would-be lover.

No pranks or practical jokes. Let your date get to know you before you pull a surprise that could backfire.

And don't exaggerate.

Money tends to be secondary to Explorers, particularly when spending money leads to adventure. But frugality can be an adventure, too. So respect a date who is careful with cash; he or she might be the one who helps to pay for your future children's education.

Don't assume that someone who doesn't instantly display a sense of humor isn't funny. Be patient. I've known people who I thought were deadly dull when I first met them, then discovered (months later!) that they were uproarious.

Explorers are not especially interested in exploring their own identity. And they can find those who engage in intense psychological analysis tiresome. But if a date wants you to talk about yourself, go with the flow. Explore virgin territory: yourself.

Most Explorers find self-disclosure difficult. But being secretive about important matters isn't considerate. If you have something in your background a date should know, be forthright. He or she will admire you for your candor and trust you more than if you withhold this information.

Most important: if you find someone you are genuinely interested in, avoid dating others and focus your energy on that person. In the right relationship, your partner will open more doors than he or she will close for you.

Finally, if you are an Explorer, think about how the other types will respond to you. Use your creativity to bond with a Negotiator. These men and women have a fine imagination and will like your originality. Be patient with a Builder's questions; help the Builder loosen up, too. Ignore competitive comments of the Director, with whom you may have some fascinating intellectual conversations. And if you meet another Explorer like yourself, revel in the excitement of finding a true play mate. But if you are interested in a long-term partnership, find out if this Explorer is also interested. If not, don't wait for him or her to change; depart.

If you are dating an Explorer, be prepared to live this romance one day at a time. And don't be dull. As the novelist Ethel Wilson summed up the Explorer's attitude, "dullness is a misdemeanor." Avoid unnecessary routines and stay flexible with your plans. Pick up on your Explorer's energy and optimism. Emphasize adventure, intellectual and physical. Explore his or her ideas. Don't smother an Explorer with details. Avoid tedious repetition. Never moralize. Be emotionally expressive. Be playful. And because Explorers don't impose limits on others, enjoy your freedom.

The Builder

"The heart is forever inexperienced," declared Henry David Thoreau. It's true. Even the wisest men and women can be fools when it comes to love. Yet each personality type has its own set of strengths and weaknesses. So it is with Builders.

Foremost, if you're a Builder, don't let your taste for plans and schedules get in the way of trying new things on a date. Be ready to change your routines—and your mind. As F. Scott Fitzgerald put it, "Vitality shows in not only the ability to persist but the ability to start over." Spontaneity and flexibility are vital when courting.

Make sure to allow for time alone with someone who interests you. Builders tend to be social. They weave elaborate networks of friends and colleagues and enjoy "keeping up" with everyone. So they are likely to bring a date along when they see their friends. But not all potential mates want to meet your crowd immediately; some would prefer to talk only to you.

Recently, for example, I had a date with a Builder who did just this. We were casual acquaintances when he invited me out. During the course of the evening he took me to three parties. I talked to his business partners at a reception for lawyers; I met his friends in Greenwich Village at a book party for someone's daughter; and I mingled at an event for Irish New Yorkers. I liked the novelty. But I never really talked with him. I found the evening tiring and unsatisfying.

Because Builders are willing to do thankless jobs that require time and energy, they often go overboard in helping others. So if you're a Builder, make sure a potential partner is interested in you before you sign up to fix his or her car or provide other time-consuming or expensive favors. Don't offer too much help too fast.

Never allow yourself to be pressured into sex. The right partner will respect your boundaries. Then, if the relationship proceeds, learn to talk about your sexual needs. Builders can sometimes be so traditional that they are reluctant to discuss what they like in bed.

Builders take the opinions of friends and family seriously. And they can be crushed by criticism. But if you're a Builder, don't let your friends' negative views of a budding romance interfere with your relationship.

Likewise, don't expect the person you have begun to date to immediately like your friends. You probably make friends easily, but for some people making friendships takes time.

Modesty can be charming, but Builders can be too modest about their achievements. Let a date know some of your accomplishments; he or she *wants* to be impressed.

Not everyone is as orderly as you are as a Builder. Overcome your perfectionism. When you meet someone who is a little messy, either overlook this trait or think of it as an opportunity to help—when the time is right.

Never be openly critical or judgmental in a love relationship.

And be careful with your moral streak. As Hannah Whitall Smith, a nineteenth-century orator, said, "The true secret of giving advice is, after you have honestly given it, to be perfectly indifferent whether it is taken or not and never persist in trying to set people right."

If you meet someone who is an expert in a field you know nothing about, ask questions. You will charm your date with your interest— and appear more interesting yourself.

Your tendency to be protective will be appreciated; just make sure you don't overdo it. You don't want to tread on your date's need for independence.

Builders like social situations to be "under control." But you don't want to appear controlling either.

Be playful. Being respectable can be admirable. Yet if you're too low-key, patient and practical, you'll be seen as boring.

Spend some money. Builders are the most frugal of the four types. But dating often requires spending. And you want to be seen as generous.

Develop your romantic side. A spontaneous word or gesture of affection can be intimate and affirming.

And take a chance. When it comes to love, one must seize the moment. "All glory comes from daring to begin," it has been said. Go for it: risk.

If you are a Builder going out with an Explorer, contain your questions about details and listen to your date's theories and ideas. Ignore a Director's tendency to make quick decisions. Make a personal connection with a Negotiator by talking about yourself, including your private feelings. And be particularly courteous with other Builders. Like you, these people have traditional values. Of all four types, another Builder is most likely to appreciate who you really are.

And if you're another type and have begun to date a Builder, remember that Builders tend to be concrete. They want to talk about the details of projects, events, social engagements and people. And they are often good at trivia. Avoid phrases like "I suspect that . . . ," "I would guess that . . ." and other indications of uncertainty. If you can, be orderly and calm. Make plans for social events and stick to the plans whenever possible. Respect the Builder's need for schedules and customs. Minimize the risks when proposing something new.

Organize your time together. Appreciate the Builder's loyalty and traditional values and remember that he or she will value the same in you.

The Director

"One ought to hold on to one's heart; for if one lets it go, one soon loses control of the head, too," wrote Nietzsche. Directors, like Builders, want to be in control—of themselves and of their romantic lives. But unlike Builders, Directors are neither cautious nor social. Most don't really enjoy dating; they would rather seal the deal and move on to other matters. So when Directors find an appropriate "mind mate," they court with determination—and speed.

I recall one Director in my past who marched into my apartment on our third date, surveyed my domain and began to measure a spot near a window in my living room. Then he remarked that this window space would be too small for his desk, his most important possession. I was appalled. I hadn't even decided whether I wanted a relationship with him, and here he was planning the details of our future life together.

If you're a Director, *go slow.*

"Patience is genius," wrote the Comte de Buffon, the French naturalist. The Director's impatience can lead to other problems while courting. These men and women like to get to the point. And they can become restless when a date's conversation meanders. Relax. Enjoy these tangents. You'll get insights and ideas, as well as learn much more about your partner.

Don't become so absorbed in your work that you neglect your search for love. Go out, even when you crave more time at your desk. Relationships are like businesses; if you want them to succeed, you must show up.

Leave your competitive and debating spirit in the office, too, so you can truly enjoy your time with a date. Not everyone likes to defend every remark with facts and logic.

Directors don't suffer fools gladly. And if they become uninterested in a date, they are likely to ignore him or her. So if you find yourself with someone who at first seems tedious, try to find *something* interesting about this individual and pursue it until you can diplomatically depart. In fact, try to listen actively by occasionally repeating what your date says. This way, he or she will know you're listening.

Although Directors make up their minds quickly, they respect self-control. Directors don't trust impulsive people. Try to enjoy these wilder beings. A "free spirit" can bring energy and creativity to the table; he or she might be brilliant, too.

Express your emotions. Directors can be overly self-contained. But most people will find you more interesting if you show some feeling, exposing your softer side.

And when a date says something emotionally revealing, don't ignore it. Acknowledge his or her feelings and show some sign of understanding.

Don't be too quick to mistrust what you're told, either. Directors can be so skeptical that they can stifle imaginative conversations.

And don't jump to conclusions about another's feelings for you. Directors are often hypersensitive to the slightest sign of rejection and misread a date's intentions.

Take care not to offer a monologue about your work, ideas or other interests. In courtship, ideas must be exchanged.

Don't take flattery seriously. Because Directors tend to say what they mean, they often assume others are equally honest and direct. But as the French saying goes, "He who can lick can bite."

Don't tease your date until you know him or her. Even then, if this person begins to get annoyed, stop. Humor is a great way to make—or break—a budding partnership.

Directors like sex and take it seriously. But they can also enjoy a one-night stand. Be careful of these brief encounters: sex can trigger all sorts of unexpected feelings in the brain—and you may find that you have accidentally jump-started a relationship you don't want or suddenly feel hurt that your mysterious partner doesn't want to see you again.

If you are a Director dating an Explorer, enjoy exploring ideas with him or her. If you're courting a Negotiator, let him or her examine all the angles of the discussion; don't push for closure. Answer the Builder's questions in detail; he or she will appreciate your fondness for exactitude. And if you are courting another Director, don't cancel dates in order to stay late at the office or to pursue your hobbies. Don't let your fellow Director cancel dates, either. Freud maintained that life has two great joys: work and love. Don't neglect the love.

Directors, like all types, communicate best with people who understand their style. So if you are dating a Director, lean back casually when sitting with him or her. This signals self-confidence, something Directors admire. Focus intently on the Director but don't lock eyes; this personality type is uncomfortable with the "anchoring gaze." Be logical, accurate and clear. Avoid small talk. Be decisive. Avoid long-winded comments or explanations; get to the point. And enjoy the Director's tendency to be forthright; with Directors you know where they stand. Don't criticize yourself; many Directors regard this as pathetic. Keep firmly in mind that Directors are often eager debaters; they like to spar, so fight back. Most important, strike up interesting conversations and pursue them in depth. Remember: Directors want a mind mate.

The Negotiator

"Every selection is a leaving aside," wrote Jean-Paul Sartre. Negotiators are poignantly aware of this. They see all the options, all the angles. And they can drive themselves crazy thinking of the endless possibilities and traps involved in starting (or ending) a relationship.

So if you are a Negotiator, stop thinking so much about the opportunities and problems you foresee down the road in a relationship.

This is not to say you shouldn't listen to your "heart." Your intuition is your brain's way of giving you valuable information. But Negotiators live in a world of nuance, where every tilt of the head, drop in tone of voice or shift of posture has meaning. Hence Negotiators can read too much into a partner's mannerisms. Simplify.

Try to be more straightforward, too. Take a tip from cats. Critic Joseph Wood Krutch wrote of them, "Cats seem to go on the principle that it never does any harm to ask for what you want." If you're a Negotiator, you can be so diplomatic that a potential partner is sometimes unclear about who you really are and what you need. Ask.

Stick up for yourself. If your date challenges you on your views, challenge back. Don't be so gracious, tactful and empathetic that you appear spineless. Fight for your point of view.

Above all, don't placate. Negotiators can be too eager to please.

Don't hold grudges either. Negotiators have a good memory for emotional experiences, particularly bad ones. And they are likely to remember minor social gaffes. But the other three types are not as skilled at handling people as most Negotiators. So try forgiving and forgetting a date's minor social transgressions.

Avoid a verbal deluge. With your excellent command of language, you can overwhelm your date with words, particularly if you are trying to impress. And when you talk, try sticking to one point at a time and mining it thoroughly before going on to the next idea. Don't develop too many sides of any issue at once.

Not everyone is interested in analyzing their identity or their motives. So avoid delving into your partner's psychological makeup or the underpinnings of the relationship unless he or she is also interested in this introspection.

Develop a healthy skepticism. Negotiators can be too trusting, sometimes appearing gullible.

Don't listen to everything your friends and family say about the person you like. You know best about what you want in a partnership.

Don't follow your partner's sexual agenda unless you are

enthusiastic about it. Set your own pace and clearly ask your partner to respect your limits.

Some people are so focused on what they are doing that they don't notice when you walk into the room. Try not to take this personally.

In fact, don't overreact to all the other little things that can go wrong in a budding partnership. Almost no one likes a drama queen or king.

When you give a compliment, make sure it's sincere. Don't give a compliment just to be polite. And when someone gives you a compliment, *accept* it. Negotiators tend to brush off compliments, thus undermining both the giver and the gift.

Don't apologize unless you are really sorry. Negotiators use apologies to make others feel comfortable. But some people regard an apology as weakness instead of graciousness.

Never multitask on a date. It makes you look as if you aren't paying attention, or even worse, that you are insincere, scatterbrained or uninterested.

Act confident, even if you aren't. Both sexes like a partner who is confident.

And don't settle for anything except a deeply meaningful, authentic relationship. Negotiators hate confrontation. And with their sense of compassion and their fear of conflict and loneliness, they are likely to remain in a bad relationship far too long. Face the trauma. Remember this Zen proverb: "The way out is through the door."

As a Negotiator, you are likely to get intimacy from face-to-face conversation about important personal matters. But it's worth remembering that other types are likely to feel intimacy under different circumstances. Explorers often feel intimate with a partner when they are *doing* something adventurous together. Builders can derive intimacy from sharing a social event with a mate, and Directors often feel close to a partner when engaging in a rigorous, sometimes competitive intellectual conversation. Find out what your partner regards as intimate, and cultivate this form of closeness as well as your own.

If you are a Negotiator who is dating an Explorer, don't take his or her charm seriously until this engaging person begins to show serious interest in you. Remember that a Builder is not asking questions to criticize. Don't be so polite with a Director that he or she thinks you're weak. Always get to the point with a Director, too. And if you are courting another Negotiator, avoid endless analysis of "the relationship."

If you are another personality type who is wooing a Negotiator, talk to your Negotiator about ideas and avoid chitchat; engage his or her thoughts and opinions. Lace your conversations with your feelings. Negotiators enjoy emotional expression. And don't hesitate to talk about yourself. Negotiators want to get to know you. In fact, they will divulge their innermost feelings in hopes that you will, too. Listen actively. Sit facing them. Lean toward them. Look straight into their eyes. Smile. And be authentic and personal; Negotiators admire warmth.

Also, remember that Negotiators aren't direct and they don't like to say "no." So read between the lines to hear what they are really saying. Avoid being competitive or aggressive with them. And don't interrupt; Negotiators take turns when talking. Above all, stimulate their imagination.

"Happiness is like a butterfly. The more you chase it, the more it will elude you. But if you turn your attention to other things, it will come and sit softly on your shoulder." Henry David Thoreau knew we can't capture happiness with a net, particularly happiness in love. But we can use what we know of human nature to prepare ourselves—so that when love alights, it comes to stay.

Here, then, are some final insights into nature's ways, data that all four personality types can use as they play the world's most ancient game, the mating game.

Be Lovable

As the Roman poet Ovid put it, "To be loved, be lovable."

Men generally seek women who are easy to talk with, warm, sweet, intelligent, energetic and self-confident, while women tend to like men who are relaxed, sensitive and smart. Both sexes also want a partner who is considerate, understanding, funny and honest. Don't try to be perfect, though. Men, as it happens, are less attracted to women they regard as perfect than to those they see as nearly perfect but who also make some blunders.

So blunder on. But be lovable.

Like Yourself

Psychologists often say that the first thing you must do to find real love is to get to know and like yourself. This can't be entirely true. For millions of years our forebears (and a lot of other creatures) found "the one" without a great deal of self-knowledge or fondness for themselves. You don't have to like yourself to find romantic love.

Nevertheless, male chimpanzees are particularly attracted to females who are self-assured. This is also true of humans. When you think you are attractive, you behave with an air of self-confidence that attracts others—as two high school students recently established.

In an intriguing scientific project in 2006, Ian Panchèvre and Stephen Odanovich approached 125 girls in malls and on the streets in San Antonio, Texas, and tried to pick them up. But as they interacted, these young researchers assumed one of five levels of self-confidence, appearing timid, hesitant, humble, self-assured or arrogant. After five minutes of conversation they told the girls about their investigation and asked them to fill out a survey of their impressions of the interaction. The self-assured pickup was the most successful courtship ploy.

In fact, self-assured people are also more friendly, open and flirtatious. So be self-assured; walk and act with poise.

And if you don't like yourself, act "as if." As a tough-minded friend of mine is fond of saying, "Get off your cross; we need the wood."

Better yet, create a phrase that you can repeat to yourself in the shower, in the car or anywhere else, something like "I love being myself because I am ————." Find something you *honestly* like about yourself and repeat it. (If you are a Negotiator, don't overthink this exercise, just try it.) But make sure this recitation puts a smile on your face, a lilt in your voice and confidence in your step.

Don't be your own worst enemy. As the poet Henry Wadsworth Longfellow put it, "Not in the clamor of the crowded street, / Not in the shouts and plaudits of the throng, / But in ourselves, are triumph and defeat."

Brag—Carefully

You might try bragging a little, too. Courting is not, foremost, about honesty; it's about winning. Every animal on the planet seems to know this except people. Do you suppose a peacock is modest about shimmering his tail feathers at passing peahens?

Shimmer your feathers.

Your ability to write a poem, play or story; your talent for composing songs or symphonies; your religious fervor, sense of humor, or moral virtue; your talent for understanding calculus; your impulse to give to charities; your political convictions; even your perseverance or kindness: all of these elaborate, seemingly unnecessary human traits evolved, at least in part, to dazzle potential mates and win the mating game.

So brag a little.

Negotiators and Builders should take special note of this advice. These types can be so modest that they are likely to underplay their assets—something Directors often regard as weak and Explorers would regard as an opportunity missed. And if you are a Director (who enjoys competition) or an Explorer (who can exaggerate), be clever about your bragging. Don't overdo it.

Don't Play "Hard to Get"

People often ask me if they should play "hard to get" at the beginning of a relationship. Frankly, I have never seen anyone succeed at this. When you become interested in someone, it's very difficult to contain your feelings. I will say, though, that playing hard to get can give you a biological advantage. Barriers intensify feelings of romantic love, what I call *frustration attraction,* probably because the brain pathways associated with pleasure, energy, focus and motivation keep working when a reward is delayed.

Nevertheless, five academic studies have investigated the payoffs of playing hard to get, and none have found that this strategy improves one's chances of winning "him" or "her." Women are less impressed with this courtship game than men. They find standoffish, overly choosy men annoying.

But I suspect each personality type gets annoyed at this courtship ploy for a different reason. Builders are dedicated to honesty, duty and loyalty; they take a high moral tone in just about all they do. So they are likely to regard someone who plays hard to get as dishonest and possibly lacking moral fiber. Moreover, Builders are pragmatic; they are looking for a help mate, and they don't respect people who construct false impediments to their plans.

Directors value their time; for them dating is a chore. So they would probably regard someone who plays hard to get as wasting their precious time. Directors are also decisive and goal-oriented. Like Builders, they are not amused by a date who adds unnecessary complications to a budding relationship.

Negotiators fall in love more regularly than the other types. So Negotiators may overlook or minimize this obstacle—particularly if they think they have found "the one." But Negotiators would be

perturbed because a potential mate who plays hard to get is not being authentic.

Only the Explorer is likely to be spurred on by the man or woman
who plays hard to get. Unpredictability excites Explorers. Moreover, Explorers are optimistic and flexible; they often find impossible situations
stimulating. But eventually even the Explorer will get bored if this game
persists.

No one really likes to be "toyed with" when looking seriously
for love.

"Come Hither" Flirting Ploys

Ever watch a pussycat flirt? Mine used to bat her would-be lover in the
face with her paw, hiss at him, strut off a few paces, sit, lick herself,
look back over her shoulder at him, then flick her tail, stand, arch her
back and thrust her butt in his direction, suggesting that he follow. He
would. She'd turn when he got near, sit, bat him again, and continue
her routine, each time letting him sit closer as he watched.

Like cats, dogs, peacocks and all other creatures, humans have
evolved an arsenal of courting behaviors. We perform these naturally,
often unconsciously.

Psychologist Monica Moore has recorded a host of these courting ploys, tactics your mother never taught you. Monica and various
research assistants sat in local pubs, restaurants and parties in St.
Louis, Missouri, over the course of two years, randomly chose 210
women and watched them flirt. Indeed, Moore discovered fifty-two
different facial, head and body cues that women use to tell men they
are interested.

The "head toss" is a common ploy. A woman throws her head
back, tilting her face quickly up, then down again. Often combined
with this is "hair flipping" (running your fingers through your hair),
"hair palming" (running your palm over your hair) or "hair stroking"
(running your whole hand through your hair several times). Some
women go "face-to-face" with a man, putting their head directly in

front of his with noses almost touching. Turtles do this, too; but they actually tap snouts. Dolphins just nibble noses.

With the "neck presentation," a woman tilts her head to the side at a forty-five-degree angle to expose her neck. Female wolves do this as well. Some women "palm," extending their upturned hand toward a suitor. Chimpanzees use this begging gesture to extend friendship. Many women also do the "solitary dance," keeping time to the music by dancing in place.

Women also "primp," smoothing their clothes. Some massage a knee; some dangle a shoe from their toe. They also touch themselves. Using what are known as *intention cues,* courting women (and some men) stroke their own arm, rub their neck, move their arm toward a potential partner on the table, slide a foot closer if they are standing, or lean in toward the other's face—behaviors they may unconsciously wish the other were doing instead. They engage in "lint picking," too, plucking a bit of fuzz from their sweater.

And women use props. As they talk, they do the "object caress," playing with keys or something on the table. Many laugh, giggle, whisper, wink or bat their eyelids to draw attention to themselves. Some stand very close to a man. Many "parade," walking with arched back, stomach in, chest out and head high as they dramatically swing their hips. And many request assistance, handing their coat to a man to help them put it on, or asking for another drink.

Many women like to play—sticking out their tongue at a man, sneaking up from behind to cover his eyes, or tickling him. And women caress men, rubbing a man's arm, stroking his thigh, hugging, patting, even playfully kissing him.

The most assertive women use the "come hither" pose, known to animal behaviorists as *lordosis.* It's probably universal among mammals. In the case of a woman, she walks away or turns away from the man, and then in one swift motion she arches her back, tips her buttocks up toward him and looks over her shoulder to gaze for a moment directly into his eyes—just like my pussycat. All other female mammals use lordosis to invite copulation. Women do it just to flirt.

Courting women, on average, used some seventy blatant flirting acts per hour, Moore reports. And those who flirt more are more regularly approached by men. Indeed, even when a woman sends a mixed message—such as smiling at a man while keeping her arms tightly wound around her body or even turned away—men still regard it as a come-on.

Men are more likely than women to believe that a person's friendly or ambiguous behavior is a sign of sexual or romantic interest.

Courting men tend to make dominant gestures toward other men, playfully punching them in the shoulder, slapping them on the back, elbowing them in the ribs. Some men stride and swagger. Some sprawl out in a chair, arms stretched across other chairs to claim more space. Men stand with their hands on their hips, or perch, stand or sit with their legs apart. They stroke their beard or chin, accentuating their masculine jaws. Some burst forth with loud laughter or grand gestures, bobbing and swaying to draw attention to themselves. And when outdoors, many pick up something and throw it, skip it, bounce it, or do something else with an object to display their strength, speed or throwing accuracy.

Face Talk

Eye language is at the core of flirting. Men and women "eye-sweep" the room, surveying it for a potential mate. They use the short, repeated, darting glance to investigate a particular individual. And they employ a prolonged gaze (three or more seconds) when they are really interested in someone. Known as the *copulatory gaze,* it's used by many different animals. Chimpanzees, for instance, stare unmercifully when they woo.

Both women and men also use the "head nod," in which the head is moved forward and back on the neck, showing they are listening.

Both sexes employ the "eyebrow flash," in which both eyebrows are raised together for about two seconds, to display their interest in another person. And both sexes accentuate their words with somewhat exaggerated arm and hand waving or with the "lip lick" or "lip pout."

Smiling is very important in flirting. I can't recommend it strongly enough, particularly to Directors, who can regard smiling as weak, and Builders, who can regard facial expressiveness as unmanly and unprotective.

We have about eighteen different kinds of smiles, only some of which we use to woo. The "simple smile" is a closed-lips gesture that people use in casual greeting to a stranger or acquaintance. You show stronger interest when you draw back your upper lip to show your upper teeth, the "upper smile." Combine this with a flash of the eyebrow (quickly up, then down) and you will send a clear signal of interest.

But for a biological reason the "open smile," with both upper and lower teeth fully exposed, is the most effective. We are built to mimic others, particularly with our faces. And when you see another person smile, you naturally pull your face into the same pose—if only for a fraction of a second. This has a powerful effect on the brain. As you move your facial muscles, you stimulate facial nerves. And these, in turn, trigger specific brain chemicals that make you feel good. This is why we feel good around happy people. As they smile, we smile, creating powerful chemical sensations of pleasure.

So if you want to make your date happy, be happy. Your smile will make you feel better—and make him or her feel better, too.

In Samoa, Papua New Guinea, Japan, Africa, up the Amazon River and in the clubs and bars of Madrid, London, New York and Buenos Aires, women also perform the "multipart flirt." First they smile at an admirer; then they lift their eyebrows quickly while opening their eyes wide to gaze straight at him; last, they drop their eyelids, tilt their head down and sideways and look away. Often a woman also covers her face or mouth with her hand as she looks askance.

Another ploy is the "coy look," a three-part gesture we must have inherited from our mammalian ancestors because possums do it, too. First you stick your neck out slightly, then you cock your head; then you look up at your suitor.

Be it a gaze, a lip lick, an eyebrow raise, a smile, or any combination of these and many other gestures, the point is to be noticed. In fact, Moore found that these signaling devices are far more effective than physical attractiveness at getting someone's attention.

Handicapping

One odd form of courtship behavior is known as *handicapping*. A man or woman will dress down, wear unattractive, ill-fitting clothes or look tousled and unkempt; some even sit where they are barely noticeable.

Recently I saw exactly this behavior in a downtown singles bar in New York. Two young women were sitting with their backs to the room, hunched over their drinks. Both were wearing baggy jeans, oversized sweaters and horrible-looking woolen ski caps. They were dreadful to behold. But when they turned to check out the action, they displayed gorgeous faces.

Days later I saw the same handicapping behavior in a friend of mine, a staggeringly handsome fellow by anybody's standards who generally wears a pin-striped suit. He signed up for a local New York dating service and felt compelled to show me the photo and written essay he had included in his entry. To my astonishment, he had submitted a picture of himself with a three-day beard, filthy hair and a truly ugly and wrinkled shirt. In his essay he wrote, "I clean up nice."

No peacock, fox, pussycat or rhino would recommend this form of courtship. Nevertheless, scientists think that many of these slobs are advertising that they are such a catch that they can win the mating game even with a huge sartorial handicap.

Loom and Crouch

There are two courtship poses we even share with fish, lobsters, lizards and frogs: "loom" and "crouch."

The loom position, also known as "standing tall," is more regularly performed by men. Take a man reading on a park bench, with shoulders hunched, stomach out and chest concave. Suddenly he sees an attractive woman strolling past. Unconsciously he straightens up, sucks in his stomach and thrusts out his chest. This "chest thrust" is a way of looking large, a sign of dominance and courtship and a part of the natural loom response. Cats bristle. Pigeons swell. Codfish bulge out their pectoral fins. Lobsters rise up onto the tips of their walking legs and strut. Male gorillas pound their pectorals. Men just arch their backs and puff out their chests.

Women shrink, or crouch. They turn in their toes, curl their shoulders and hang their heads. The message: "I'm harmless; come hither and defend me." Many other creatures crouch to signal friendly deference. Wolves and coyotes tuck their tails between their legs and slink. Lizards and toads move their whole bodies up and down. And many species bow. Chimps nod their heads so rapidly and repeatedly that primatologists call it *bobbing*.

Men and women swell and shrink in countries around the world—signaling their availability and interest. So if a would-be partner office colleague expresses any of these postures and gestures, even fleetingly, you probably have someone courting you.

The Five-Part Pickup

Anthropologist David Givens and biologist Tim Perper have, like Monica Moore, spent hundreds of hours in American singles bars watching men and women pick each other up. Both found that most pickups follow a specific pattern—a blueprint that all personality types are likely to use when they court—involving five steps:

Step one: Men and women establish a territory—a bar stool, chair, bench or place to lean, somewhere they can see and be seen.

Next: They begin to draw attention to themselves. The loud laugh. The dramatic arm gesture. Swaying, stretching, dancing in place. Swaggering or clomping back and forth to the jukebox, bathroom or other nearby spot. Adjusting their sweater. Running their fingers through their hair. Tugging at their chin. Picking the lint off their jacket, "self-grooming." Breaking away from friends to stand at the bar alone. We draw attention to ourselves in these and many other ways, all to say, "I'm here."

Talking is the third phase of the pickup. Women often adopt the high, mellifluous tones that appeal to men, while men unconsciously keep their voice low and deep, a turn-on for women. Regardless, the best leads are either compliments or questions because the other party must respond. In fact, unexpected praise can put a man in a "good mood" and make him more friendly toward an unfamiliar woman.

Touching comes next in this primordial dance. Generally the woman touches first, casually grazing a man's shoulder, arm or wrist as she asks, "How's the food here?" or "Do you have the time?" If he flinches or otherwise withdraws, she won't try again. If he does nothing, she may venture another casual touch. And if he casually touches back, they may both begin to nonchalantly touch each other now and then.

Last, step five: "keeping time," also known as *interactional synchrony.* When two people begin to "click," they pivot toward each other, swiveling until their shoulders are almost aligned. Then they synchronize their movements. As he picks up his beer, she picks up hers. As she crosses her legs, he crosses his. At first, they may "keep time" only briefly. Then they move in tandem more often—until they are moving in perfect rhythm with each other, to a single beat. When two people mimic each other's movements, even their brain waves get "in sync."

No wonder so many singles clubs and bars have rhythmic music and a dance floor. As men and women keep the beat, they naturally get into body synchrony with those around them—escalating the courting process.

"Keeping time" can be particularly effective. I tried this once on a luncheon date. I had been utterly unsuccessful in my past attempts to move this friendship toward romance. So I mimicked his every move. When he ordered steak and wine, I ordered steak and wine. Every time he sawed at his meat, I sawed at mine. Every time he sipped his drink, I sipped mine. I cocked my head when he cocked his. I even adopted his pace of conversation, talking slowly.

It worked.

Active Listening and Humor

"Listen or thy tongue will keep thee deaf." These words of wisdom are an American Indian proverb. To build intimacy with a date, you must listen—actively. Move your body; nod your head; smile when your date is smiling: whatever you do, get connected, intellectually, emotionally and physically. And if you find your mind meandering, lean in toward him or her. Look carefully as this person speaks. Repeat some of their phrases. Add your comments. Courtship runs on give-and-take; messages must be returned. You must listen and respond.

And if you can possibly be funny, do.

Different types are likely to gravitate to different forms of humor. Explorers tend to like outrageous antics and non sequiturs that have surprise and action. Builders are often fond of trivia and anecdotes about people, places and things. Directors more often use puns and paradoxes that structurally appose two or more independent elements. And Negotiators tend to like stories about people.

But whatever humor you can produce, produce it. Wit is a powerful courting tool as well as good medicine. It draws a couple together, reduces stress and stimulates brain regions associated with pleasure and reward.

Who Pays for Dinner?

You are at a bar, a club, a restaurant or coffee shop. You have just shared some food and drink with a person you met last week. It's your first real date. And it's not clear who invited whom. Who pays?

We instinctively know this is an important issue, for it has a long evolutionary history. Throughout the animal kingdom males feed females and expect sex in return. A male chimp offers a female a bit of sugarcane or meat; the female consumes her prize; then she turns and copulates with him. A male tern feeds a female a small fish in return for sex. "Courtship feeding" is so common in the animal kingdom, and so regularly linked with sex, that when applied to insects it's called the "nuptial gift."

The exchange of food for sex is commonplace among peoples around the world as well. A piece of fish, a slab of beef, a box of chocolates: men provide. And where they can't give food, they offer beer, tobacco, jewelry, cloth or flowers. And this offering is regularly part of the courting process.

So be careful when you offer food. And think before you take food or other gifts. Deep in the brains of both the giver and receiver, an important exchange is being made—an exchange that often leads to sex and love.

But if you're interested in someone, buying dinner for them is a good gift. It's costly and thus a signal of your financial status and generosity. Moreover, it has no long-term value: it must be consumed on the spot—and you get to share the fun.

Kissing

Kissing has special effects on the brain, as you undoubtedly know.

Kissing stimulates *all* the senses. When you kiss, you see your partner, as well as smell, taste, hear and feel him or her. The lips have a huge number of neurons that pick up the slightest messages. Then five of the twelve cranial nerves escort these intense sensations to the brain, where they are processed in an impressive factory, the so-matosensory cortex. The vast majority of this brain region is devoted to processing sensations from around the nose and mouth—giving your lips and tongue their exquisite sensitivity.

Kissing affects more than just your senses, too. It boosts your pulse and blood pressure, dilates your pupils and deepens breathing. Kissing raises levels of oxytocin, the chemical associated with trust and attachment. Kissing also reduces the stress hormone, cortisol. So kissing bombards the brain, bringing pleasure and relaxation.

It also gives you vital information. With this odd habit, we pick up signals about what our partner has been eating, drinking and smoking, as well as essential data about his or her genetic makeup. Moreover, saliva contains information about certain genes in the immune system. We are naturally drawn to those with a slightly different immune system—most likely to produce genetic variety in our young. And as you kiss your partner, you unconsciously pick up these crucial data. The smell of a woman's breath and saliva changes across the menstrual cycle, too, broadcasting the extent of her fertility.

Women report that they kiss a new partner primarily to collect information. Men, however, use kissing differently. With their smooching, they hope to woo a woman into bed. Their hopes are founded in biology. A man's saliva is loaded with testosterone, the hormone that triggers sexual craving. Perhaps this is why men like wetter kisses than women do. The sloppier the kiss, the more testosterone the man transmits.

Both men and women say that kissing brings them emotionally closer to their partner—at least a good kiss does. Tender; passionate; considerate: people tend to reason that a sensitive kisser will be a suitable partner, spouse and co-parent. However, an unpleasant first kiss can kill a budding relationship. In a recent study of 58 men and 122 women, 59 percent of men and 66 percent of women said they had ended a romance after the first kiss. It was, for them, the kiss of death.

Over 90 percent of people around the world kiss. And in cultures where men and women find kissing disgusting, lovers lick, rub, suck, nip and blow on each other's faces prior to copulation. All kinds of other creatures kiss also or caress around the mouth. Bonobos, our closest chimp relatives, kiss with a deep tongue kiss. Dogs lick. Moles rub noses. Elephants put their trunks into one another's mouths. Albatross tap their bills together. Kissing is natural.

But because a kiss can start (or end) a partnership, think before you kiss.

Casual Sex?

Sex has many health benefits. Frequent sex improves your sense of smell, reduces the risk of heart disease, improves overall fitness, contributes to weight loss, relieves pain, reduces the frequency of colds and flu and improves bladder control. Having good sex also makes you want to have more sex, in large part because sexual activity elevates levels of testosterone.

Having sex can also reduce depression, particularly in women because seminal fluid is a cocktail of potent chemicals. The dopamine and norepinephrine in this male liquor relieve stress and give energy, optimism, focus and motivation. Oxytocin and vasopressin in seminal fluid can give feelings of deep attachment as well. And the follicle stimulating hormone (FSH) and luteinizing hormone (LH) found in male ejaculate regulate a woman's monthly menstrual cycle.

In short, sex with the right person at the right time is good for you.

But what about casual sex with someone you just met?

I'm not in the "should" business; I don't wish to tell you how to conduct your life. But here is one of Mother Nature's secrets: Casual sex is *rarely* casual. Fondling the genitals stimulates the production of dopamine—the brain chemical associated with feelings of intense romantic love. Just after orgasm, you get a spurt of norepinephrine too, the closely related chemical associated with energy, exhilaration and focused attention. So when you have sex with someone you hardly know, you can stimulate these powerful neural juices, pushing you toward feelings of passionate romantic love. Moreover, with orgasm you experience a flood of oxytocin and vasopressin, the "cuddle chemicals" associated with attachment. Hence you can begin to experience deep feelings of union with this partner, too.

In fact, I have often thought that men and women unconsciously "hook up" with a new date precisely for this purpose—to trigger the brain circuits for romance and attachment in themselves and a partner and thus jump-start a new relationship.

Interestingly, anthropologist Justin Garcia and others have recently come to the same conclusion. In his study of casual sex on an American college campus, Garcia found that 50 percent of women and 52 percent of men who initiated a "one-night stand" were eager to jump-start a longer connection with this partner and one-third of these "hook ups" turned into romantic relationships. Moreover, when this strategy failed to trigger a relationship, one of the partners regularly became depressed—suggesting that this individual had hoped for a longer, more meaningful connection.

Casual sex can lead to love. So think before you step into its deep waters.

Commitment

Last, if you're a man, don't take every minor gesture a woman makes as a sign of her sexual interest in you. And if you're a woman, don't assume that men aren't interested in commitment. They are. Men fall in love faster than women do, most likely because they are so visual. They are often more dependent on their partnerships because they make fewer close male friends with whom they can share their feelings. And divorced and separated men tend to kill themselves two and a half times more often than their married counterparts, whereas divorced women are not more likely to commit suicide.

If men overestimate a woman's sexual interest, women *underestimate* a man's willingness to commit.

Born to Love

"Advice is like snow; the softer it falls, the longer it dwells upon, and the deeper it sinks into the mind," wrote the poet Samuel Taylor Coleridge. Who knows which of the above suggestions will sink into your mind and help you find and keep "the one"? Perhaps some day you will write and tell me. But of one thing I am quite certain: you *will* fall in love.

We are born to love. Romantic love is a human drive, one of three basic brain systems that evolved millions of years ago. The *sex drive* motivates us to seek sex with a range of partners; *romantic love* predisposes us to focus our mating energy on just one individual at a time; and feelings of deep *attachment* inspire us to stay with a partner long enough to raise our children as a team.

But of these three neural circuits, romantic love is often the most powerful. It is certainly more powerful than the sex drive. After all, if you ask a friend to have sex with you and he or she declines, you don't kill yourself. But around the world, some men and women who have been rejected in love commit suicide or slip into severe depression. Some even die of a broken heart, caused by a stroke or heart attack brought on by stress.

"The human heart is like a ship on a stormy sea driven about by winds blowing from all four corners of heaven," wrote Martin Luther. We love—in spite of every other force on earth. Patrick and Suzanne, whose wedding I attended; George and Martha Washington; Charles and Emma Darwin; King Hasaw Chan K'awil, whose tomb has touched his beloved's grave with its shadow for some 1,300 years: men and women loved a million years ago; men and women will love a million years from now.

And romance can last for many years—as a new brain-scanning experiment has shown. Psychologists Bianca Acevedo and Art Aron, neuroscientist Lucy Brown and I have just completed this study, in which we scanned the brains of men and women who reported that they were still in love after an average of twenty one years of marriage.

These participants showed activity in brain regions associated with all three basic human drives: the sex drive, romantic passion and attachment to their beloved.

Love can endure—bringing a lifetime of laughter and adventure, fulfilling sex, fascinating conversations, warm experiences with children, kin and friends and lasting feelings of passion and union with another.

But you have to pick the right partner. Perhaps you are an Explorer who is attracted to a play mate, a Builder who wants to find a help mate, a Director who will choose a mind mate or a Negotiator who gravitates to a soul mate. Regardless of whom you seek, you have eavesdropped on Mother Nature. You know more about the hand that you've been dealt. And you have learned a number of Nature's secrets about romance.

Use these secrets. There will always be magic to love. But knowledge is power. If you know who you are, what you seek and how you and others love, you can capture that magic, find and keep real love, and make your dreams come true.

NOTES

I have not referenced any of the academic studies or other source material that support the individual facts in this book. Instead, I have included an extensive bibliography that lists all the supporting materials I consulted during my research and writing. However, I am currently writing a series of academic articles on these data and ideas that will cite all scientific references. If you would like to know the supporting material for any specific fact in this book, please e-mail me. Also, many of the names of people in the book have been changed.

The following notes elaborate on specific points.

Chapter 1: EAVESDROPPING ON MOTHER NATURE

1. Psychologists have long maintained that personality traits interact and covary in specific, complex, hierarchical and additive patterns (Goldberg 1993; Costa and McCrae 1995; Reif and Lesch 2003), and they systematize these interrelationships using various terms, concepts and models of personality structure

including *first-order, second-order* and *third-order* factors; *dimensions* of personality; *domains* of personality and the *facets* of these domains; *sets, subsets* and *elements* of subsets of personality; *circumplexes;* and *discrete clusters of covarying traits* (Costa and McCrae 1995). These data and theories are beyond the scope of this book. Moreover, with the exception of factor analysis, I have not attempted to systematize the hierarchical, lateral and overlapping relationships between the constellations of traits associated with my four personality types. I have also excluded data on *genomics,* which analyzes the quantitative trait loci (QTL) that produce complex behaviors, and *behavioral genomics,* which studies how these multiple-gene systems affect behavior. I focus only on traits with known genetic underpinnings that contribute specifically to the four behavior syndromes discussed in this book, as well as traits that appear to covary with these genetically based traits.

2. Scientists generally regard the word *type* as inappropriate because it refers to categories rather than *dimensions* or *scales,* terms commonly used by psychologists, or *behavior syndromes,* the term used by behavior geneticists. In this book I use *types* primarily for literary purposes.

3. The activity of and interactions between chemical systems are governed by many factors, including the amount of production of the neurotransmitter or hormone; the enzymes controlling the production of each neurochemical; the sensitivity and/or number of postsynaptic receptors that receive the chemical; the sensitivity of presynaptic receptors that regulate the production of the neurochemical through a negative feedback system; and catabolism mechanisms regulated by enzymes such as MAO and others. Moreover, different combinations of neurochemicals can produce the same net behavior in different people. Although the full biological profile for each behavior constellation discussed in this book is unknown, the book is based on data that indicate that these biological constellations are significantly different from one another and contribute to who we are.

4. While related chemical systems often act similarly or in tandem or trigger one another, these same brain systems can also suppress one another or other brain systems—altering ratios, interactions and behavior. In some cases the behaviors discussed in this book are associated with decreased activity in a system rather than elevated activity. Behavior also varies according to which of several different alleles (genetic variations) are expressed in each system. I do not discuss GABA, glutamate, the opioids,

prolactin, vasopressin or other neural systems likely to affect these behavior constellations.

5. Regarding this test in the American population, the Cronback's coefficient alpha of reliability is: Explorer r=0.791; Builder r=0.793; Director r=0.809; Negotiator r=0.783. These scores are based on a sample of 39,913 members of Chemistry.com, and they are well within the range of meaningful reliability. Using factor analysis, each primary scale breaks down to three subscales: (a) Explorer: novelty seeking; energetic; restless; (b) Builder: moral and traditional; meticulous and cautious; plans carefully; (c) Director: intellectual; mechanically skilled; straightforward/tough-minded; (d) Negotiator: imaginative; emotionally intense; seeks intimacy. The Cronback's alpha has also been calculated for Australia: Exp: 0.800; Bui: 0.806; Dir: 0.818; Neg: 0.796.

6. Several validity measures illustrate that these four temperament types represent, respectively, the activity of central dopamine, serotonin, testosterone and estrogen. For example, the ratio of the length of the second finger to that of the fourth finger reflects testosterone and estrogen exposure in the womb (Manning 2002); and as predicted, the Director scale correlates positively with a longer fourth digit, indicating elevated exposure to fetal testosterone, and the Negotiator scale correlates positively with a longer or equally long second digit, indicating elevated exposure to estrogen in the womb. Moreover, in the sample of 39,913 members of Chemistry.com, significantly more women scored high on the Negotiator (estrogen) scale, while significantly more men scored high on the Director (testosterone) scale. The Builder scale correlated positively with religiosity, a trait associated with the serotonin system. And the Explorer scale correlated positively with several measures for novelty seeking, a trait in the dopamine system. See Fisher, Rich, Island and Marchalik (in preparation) for validity data on several specific measures. Last, with Heide Island, Jonathan Rich and David Zava, I am administering this test to two hundred subjects (not associated with Chemistry.com) from whom we also take samples of bodily fluids to establish hormone and neurotransmitter activity. I have also begun genetic testing (also using subjects not associated with Chemistry.com) to correlate these questions with specific biological profiles.

7. These data are supported by previous data on mate choice. Previous studies indicate that individuals with significant dopamine expression (Explorers) tend to marry one another (Zuckerman 1994), and that those who are cautious (a core trait of the Builder) also marry one another (Zuckerman 1994).

Studies also confirm that very feminine, high-estrogen women (Negotiators) have a stronger preference for very masculine, high-testosterone men (Directors) than do women who express less estrogen (Feingold et al. 2005); and high-testosterone men (Directors) gravitate to high-estrogen women (Negotiators) (Jones 1995). All of these studies support the results of the Mate Choice Study: that Explorers and Builders gravitate to individuals who are similar in temperament, while Directors and Negotiators are particularly drawn to individuals who complement them in temperament.

8. Mike Was did an informal study of word usage on a sample of politicians, comedians and others on television, using this same system of personality types. He found that each type uses a constellation of words that correspond well with the words each type uses in courtship. See Was et al. 1999.

Chapter 3: THE HAND YOU'RE DEALT

1. An example of how the environment can tailor genetics is the baby rhesus monkey (Suomi 2005). Rhesus monkeys inherit one of two versions (alleles) of a particular gene in the serotonin system. Those with the short version tend to express extreme impulsivity and inappropriate aggression as they grow up. But if they are removed from their mother and placed with a foster mother who is stable, attentive and nurturing, they grow up to be normal instead. Good mothering can triumph over a bad gene.

2. A highly respected academic test, the NEO-PI or Neuroticism-Extroversion-Openness Personality Inventory, measures five specific personality constellations that have been identified in humans worldwide and in many other creatures (Gosling 2001; McCrae and Costa 1997; McCrae 2002; King and Figueredo 1997). Known as the "Big Five," these five trait constellations can be remembered by the acronym OCEAN: Openness to new experiences; Conscientiousness; Extroversion; Agreeableness; and Neuroticism (or Anxiety) (Gosling 2001; McCrae and Costa 1997; McCrae and Costa 1987; McCrae 2002). All five trait constellations are inherited (Costa and McCrae 1992; Bouchard 1994; Jang et al. 1998; Loehlin et al. 1998; McCrae et al. 2005; Allik and McCrae 2004). And several correspond well with the four trait constellations discussed in this book: openness to new experience is a core trait of the Explorer; conscientiousness is central to the Builder; and agreeableness is a primary trait of the Negotiator. Nevertheless, the NEO-PI breaks down each of the Big Five personality factors

into several facets, some of which do not correspond. For example, one aspect of agreeableness on the NEO-PI scale is straightforwardness. Straightforwardness has not been associated with estrogen, the Negotiator. Some facets of openness (such as imagination) apply more to the Negotiator than to the Explorer. However, many traits in the scale for openness apply directly to the Explorer, including open-mindedness, unconventionality, tolerance and creativity (Neyer and Voigt 2004). Tests of psychological traits differ, depending in part on how researchers define the traits they are studying.

3. The Myers-Briggs Type Indicator (MBTI) measures your preferences on four scales: The Sensing-iNtuition scale (S-N) measures your preference for what is observable and practical as opposed to your penchant for the abstract and theoretical. The Thinking-Feeling scale (T-F) measures how much you value logical analysis as opposed to affiliation and feeling. The Judging-Perceiving scale (J-P) measures your taste for order and rules as opposed to flexibility and spontaneity. And the Extroversion-Introversion scale (E-I) measures your preference for the external world of others versus your inner universe of ideas. Hence you will be one of sixteen types, depending on your combination of preferences on all four scales. I believe the "Perceiving" type is actually the dopamine-rich Explorer; the "Sensing" and "Judging" types both describe the serotonin-rich Builder; the "Thinking" type is the Director; and the "iNtuitive" and "Feeling" types describe the high-estrogen Negotiator. The Myers-Briggs Type Indicator is a controversial test in academic circles. Its basic scales meet academic standards for internal consistency, validity and test-retest reliability (Myers et al. 1998; Hunsley et al. 2003; Capraro and Capraro 2002). For example, Form M coefficient alpha results for a recent sample of 2,859 male and female adults were: Extroversion-Introversion scale, 0.91; Sensing-iNtuition scale, 0.92; Thinking-Feeling scale, 0.91; Judging-Perceiving scale, 0.92 (Myers et al. 1998). Yet many scientists regard it as flawed in other ways (Lorr 1991; Pittenger 2005; Bess and Harvey 2002; Boyle 1995). They question the reliability and validity of the sixteen types; the cutoffs used to assign individuals to a particular type; the bimodality of scores on each scale; the lack of recognition that these traits are expressed in continuous degrees, hence they are quantitative rather than qualitative traits; the use of a "forced choice" instrument; the practice of converting scale scores to type categories; variability of test-retest scores; the influence of age-related factors on test scores; and lack of proof that this measure can accurately predict appropriate educational, career and employment choices, or be applied to employee selection, job performance, interpersonal compatibility or

other applications (Hunsley et al. 2003; Furnham et al. 2003; Pittenger 2005; Zimmerman et al. 2001; Bess and Harvey 2002).

Chapter 4: DRINK LIFE TO THE LEES

1. Three of the most researched personality tests, the NEO-PI (chapter 3, footnote 2), the MBTI (chapter 3, footnote 3) and the TPQ, have recorded aspects of this personality scale, the Explorer. The NEO-PI refers to this factor as "openness to experience." It reports that these individuals are intellectually curious, have a wide range of interests and are open to new ideas, new theories and new intellectual and/or physical adventures. These people are flexible, adaptable, tolerant, open-minded, liberal, spontaneous, imaginative and emotional (NEO PI-R Form S; Gosling et al. 2003; McCrae and Costa 1990; Zentner 2005; McCrae 1987). The MBTI Form M reports that these men and women (the Perceiving type) collect facts, data and ideas (Quenk 2000); they are original, unconventional, curious and creative (Keirsey 1998). They also gravitate to words like unplanned, casual, unconstrained and easygoing and want to "go with the flow," do things "on the spur of the moment" and "be free to do whatever looks like fun" (MBTI Form M). The Tridimensional Personality Questionnaire (TPQ) (Cloninger 1987) measures three major, heritable personality dimensions: harm avoidance, reward dependence and novelty seeking. Novelty seeking is a core trait of the Explorer, while harm avoidance is a core trait of the Builder.

2. Two enzymes are involved in sensation seeking, remembered by their acronyms COMT and MAO; these enzymes break down dopamine, norepinephrine and serotonin. A tiny variation in the COMT gene is associated with seeking novelty (Drabant et al. 2006). MAO comes in two varieties: MAOa regulates norepinephrine and serotonin; MAOb controls dopamine. Variations in both systems can contribute to sensation seeking (Zuckerman 1994; Johansson et al. 1983; Meyer-Lindenberg et al. 2006; Sostek et al. 1981). Elevated activity of testosterone may play a role in sensation seeking because testosterone lowers MAO levels in the brain, thereby *raising* the impact of dopamine (Zuckerman 1994). Testosterone contributes to some aspects of novelty seeking, particularly the craving for action adventure and unusual sexual experiences, wild parties and illegal drugs (Zuckerman 1994). Sensation seekers have *lower* levels of norepinephrine. Norepinephrine activates when you experience any kind of novelty, enabling you to focus your attention, and sensation seekers may seek arousing stimuli to trigger this chemical system

and thus alleviate monotony (Zuckerman 1994). Sensation seeking and impulsivity both have been associated with an *underactive* serotonin system as well (Zuckerman 1994; Manuck et al. 2000). This is to be expected because elevated dopamine activity suppresses serotonin circuits and serotonin activity suppresses dopamine (Stahl 2000).

3. Focus is also associated with norepinephrine and testosterone, although the role of testosterone is most likely the result of activity in the dopamine system. Testosterone and dopamine have a positive correlation: elevated activity in each system elevates activity in the other.

4. Data on ADHD (attention-deficit / hyperactivity disorder) and ADD (attention deficit disorder) suggest that these genetically based syndromes associated with impulsivity, hyperactivity and hypersensitive sensory responses are correlated with *reduced* activity in parts of the dopamine system, particularly in regions of the prefrontal cortex (Zentall and Meyer 1987; Zuckerman 1994:358). Paradoxically, psychostimulant drugs relieve this hyperactivity. Neuroscientists regard this as a paradox because high concentrations of dopamine generally *promote* energy and activity (Marx 1999). Why does raising dopamine concentrations in hyperactive people calm them, while raising dopamine concentrations in normal people makes them hyperactive? Some scientists have suggested that low doses of psychostimulants do not affect the dopamine system but instead boost serotonin to balance the high dopamine activity, producing calming effects (Marx 1999). However, serotonin boosters, such as Prozac, do not generally calm ADHD and ADD patients. Other scientists believe psychostimulants raise levels of dopamine so high that dopamine neurons become habituated to them (Marx 1999). Still others suspect these drugs create a slow rise of dopamine activity, creating a different affect than a fast rise. I suspect that the dopamine system has a threshold effect: individuals both below and above this threshold are hyperactive. I suggest this because other neurochemicals, including norepinephrine, serotonin and oxytocin affect behavior differently in different concentrations and in different ratios with other chemical systems; because scientists believe that ADHD and ADD are caused by defects in several gene systems (Gainetdinov et al. 1999); and because different alleles in each of these systems are associated with different degrees of novelty seeking and other Explorer behaviors.

5. Dopamine inhibits the secretion of prolactin (De Zegher et al. 1993), a chemical associated with agreeableness. Perhaps dopamine's suppression of

prolactin enables the Explorer to be less agreeable, and thus more autonomous in his or her thinking and deeds.

Chapter 5: A PILLAR OF SOCIETY

1. Serotonin is made in several parts of the body. But the serotonin that affects
temperament is made in the raphe nucleus in the brain. Serotonin is received
in at least fourteen types of receptor sites that influence many bodily pro-
cesses, including body temperature, cardiovascular activity, respiration, sleeping, eating, aggression and aspects of personality.

2. Low serotonin activity is associated with aggressiveness, impulsivity, anger
and hostility (Manuck et al. 1998; Sundblad and Eriksson 1997; Davidge et al.
2004). Two alleles of a gene in the serotonin system contribute to this bel-
ligerence (Suomi 2005). Rhesus monkeys who inherit the short version have
less serotonin (and more testosterone) and are more likely to be aggressive,
impulsive and alcoholic. Some 5 to 10 percent of rhesus monkeys inherit this
short version, and many die young because they take deadly risks in fights
(Higley et al. 1996; Higley and Suomi 1989). But Steve Suomi believes this
gene has been maintained in rhesus monkeys because these impulsive, bel-
ligerent creatures are also more exploratory—an essential quality in a species
that, like humans, must survive in harsh, changeable environments. Serotonin
and testosterone have a negative relationship. But testosterone does not always
suppress serotonin. Testosterone breaks down into androgenic and estrogenic
metabolites. When testosterone converts to estrogen, it elevates serotonin activity (Sumner and Fink 1998), reducing aggressive behaviors (Simon et al.
1998).

3. Genetic variations in the serotonin transporter system (5-HTTLPR) produce individual variations in anxiety, harm avoidance and traditionalism (Barr
et al. 2003; Golimbet et al. 2004; Parks et al. 1998). There are two alleles at this
locus, genotype *ss* and genotype *ll*. Carriers of the *ss* type are more cautious
and more likely to observe social norms than are those of the *ll* or the *ls* genotypes (Golimbet et al. 2003). Those with the *ss* genotype are also more anxious
(Murakami et al. 1999). Harm avoidance and self-directedness are also associated with genetic variations in other aspects of the serotonin system (Peirson
et al. 1999; Melke et al. 2003).

4. Polymorphic alleles of the serotonin 2A receptor gene (5-HTR2A), T102C
and A1438G, are associated with variations in emotionality, activity and

sociability. Individuals with one variation in the T102C gene display lower levels of anxiety, have higher scores on a scale of hypomania (mild mania) and are more extroverted (Golimbet et al. 2004). Those with the A1438G heterozygote A/G (rather than the homozygote G/G) are also more extroverted and have lower scores on a scale of "No Close Friends" (Golimbet et al. 2004). Selective serotonin reuptake inhibitors (SSRIs) increase social affiliation in correlation with elevated plasma levels of these drugs (Knutson et al. 1998).

5. A PET scan study of the 5-HT1A receptor density in the dorsal raphe nuclei, hippocampal formation and neocortex in fifteen normal male subjects showed a correlation between self-transcendence and serotonin activity (Borg et al. 2003). Borg and colleagues argue that variability in the 5-HT1A receptor density may explain why people vary in their religious and/or spiritual fervor. Self-transcendence (or religiosity) is also associated with a serotonin transporter SLC6A4 gene polymorphism (Bachner-Melman et al. 2005). Other gene variants are also associated with self-transcendence or religiosity (Golimbet et al. 2004; Ham et al. 2004).

Chapter 6: ALWAYS TO THE STARS

1. The environment in the womb can change the balance of these fetal hormones. Some males are exposed to more estrogen in the womb, coming from the placenta and their mother's blood; some females are exposed to more testosterone from their mother's adrenal glands. These phenomena affect digit ratio and the degree to which one expresses personality traits associated with testosterone and estrogen.

2. During their lifetime, women produce small amounts of testosterone in their ovaries and adrenal glands, while men produce small amounts of estrogen in their testes and fat cells. With the enzyme aromatase, men also convert testosterone into estrogen. The bodily ratios of testosterone and estrogen change with daily, monthly and life-cycle rhythms. Puberty, the aging process and menopause all affect one's bodily ratios of estrogen and testosterone. Cultural circumstances also affect this balance. For example, testosterone is decreased in a man when he enters a pair-bond with a woman (Shur et al. 2008). Moreover, paternal behavior elevates estrogen (aromatized from testosterone and/or produced in the testes), which in turn stimulates the production of prolactin in the pituitary and thus decreases the activity of dopamine and testosterone in fathers.

3. Girls exposed to high levels of male hormones in the womb (a syndrome known as congenital adrenal hypoplasia, CAH) prefer to play with cars and trucks, and they score in the male range on tests of math and spatial skills (Berenbaum and Hines 1992; Resnick et al. 1986; see Boulton and Smith 1992). A girl with a twin brother is also more likely to be tomboyish, because her brother's fetal testosterone can increase her own testosterone receptors.

Chapter 7: THE PHILOSOPHER KING

1. A specific gene or gene cluster on the X chromosome (associated with estrogen production) contributes to this brain architecture. This bit of DNA influences the formation of several regions in the prefrontal cortex that play a role in the abilities to accumulate data, assemble it into patterns, weigh the relevance of this information and use these data to make decisions (Holyoak and Kroger 1995; Grafman 1995; Boller et al. 1995). This gene contributes to several cognitive traits, including the ability to pick up and integrate the nuances of social interactions, a contextual skill (Skuse et al. 1997; Tucker et al. 1995).

2. Many chemicals can affect the activities of and ratio between estrogen and testosterone. For example, sex hormone binding globulin (SHBG) binds to these hormones, making them unable to act on receptors; so although the levels of testosterone and/or estrogen can be high, these high levels may not indicate the effectiveness of these hormones. Only free (unbound) testosterone and estrogen can affect receptors and express themselves. Moreover, increasing levels of estrogen increases SHBG (a and d), which *decreases* the bioavailability of testosterone.

3. A brain region that may be involved in intuition is the amygdala, a small factory in limbic circuits associated with registering and generating the emotions (LeDoux 1996). Researchers put subjects into a brain-scanning machine (PET) and showed them photos of angry faces. When the participants saw and recognized these furious expressions, their *left* amygdala became active. But when the angry face was flashed too quickly to be consciously seen, the right amygdala reacted instead (Mlot 1998). Individuals with well-connected brains may excel at picking up these subconscious feelings and integrating them in the conscious mind—thus collecting data that many would attribute to intuition.

4. The drive to connect has genetic payoffs and evolutionary roots. Female pygmy chimps with the "right" connections acquire more food and social support (Parish 1994) and bear their infants in faster succession (Smuts 1997). And the children of well-connected females mature faster to bear infants of their own.

5. Empathy activates brain regions that integrate data from *multiple* sources (Ochsner et al. 2004), suggesting that those with more long-range connectivity are likely to be more empathetic (Baron-Cohen 2003; Baron-Cohen et al. 2005). Baron-Cohen (2003) believes that Empathizers pay attention to the context of the interaction in order to absorb vital cues about another's mental state and then figure out how to act appropriately.

6. The right hemisphere plays a larger role than the left hemisphere in reading facial expressions and detecting emotions in one's tone of voice (Tucker et al. 1995; Safer 1981). But we interpret these nonverbal data in centers of the left cerebral cortex. So those with greater integration between the hemispheres are better equipped to interpret the nonverbal data they have collected (Safer 1981).

Chapter 8: PLAY MATES, HELP MATES, MIND MATES, SOUL MATES

1. The terms *play mate, help mate, mind mate* and *soul mate* are adopted from the personality typology of the psychologist David Keirsey (1998).

2. Richard Ebstein (2006) found a gene in the dopamine system linked with heightened sexual arousal. He and his colleagues asked 148 college students to answer a series of questions about sex, among them "How important is sex in your life?" and "How often do you have sexual fantasies?" Then he matched each student's questionnaire scores with several of their genetic traits, including the DRD4 gene in the dopamine system. Those with a particular variant of this gene scored significantly higher in sexual desire than others. Some 20 percent of Ebstein's population carried this genetic variant.

Chapter 9: FALLING IN LOVE

1. When scientists recorded the brain activity of young heterosexual men (ages twenty-one to thirty-five) as they looked at women with beautiful faces, the left ventral tegmental area became active (Aharon et al. 2001); this brain

region is associated with focus, energy and often pleasure. Perhaps this is why people like to look at celebrities: attractive faces stimulate the brain.

2. Studies in five cultures indicate that men are drawn to two kinds of women: those with large eyes, small noses, full lips and small chins (Jones 1995), signs of estrogen; and those with high cheekbones, high foreheads, large pupils and a big smile, signs of testosterone—they regard the latter as sexy women (Cunningham et al. 1986). I suspect that if they redid these studies using my Personality Type Test, they would find that high-testosterone men preferred the more feminine faces, while high-estrogen men preferred the sexier (higher-testosterone) female faces.

3. Women gravitate to men with chiseled cheekbones and rugged jaws (Buss and Schmitt 1993; Kenrick et al. 1993; Gangestad and Thornhill 1997), signs of testosterone. However, British and Japanese women prefer more masculine male faces only around the time of ovulation; at other times of the menstrual cycle they prefer softer, more feminine, boyish faces (Johnston 1999). So biologist Randy Thornhill believes that women have two natural preferences: As they become ripe to conceive, women seek men with good genes. At other times of the menstrual cycle, they favor men who display signs of commitment. Interestingly, women *without* a partner preferred more feminine male faces even at ovulation—probably because they were seeking a long-term mate.

4. This male appetite makes genetic sense. A woman's waist-to-hip ratio is largely produced by genes; and those whose waist circumference is about 70 percent of their hip circumference have the right balance of estrogen, testosterone and other hormones to produce healthy babies. Women who vary considerably from these proportions find it harder to get pregnant and have more miscarriages and more chronic diseases, such as diabetes, hypertension, heart disease, certain cancers, circulation problems and various personality disorders (Singh 1993; Singh 2002).

5. Studies with sheep support this Freudian hypothesis. Neuroscientist Keith Kendrick has "cross fostered" baby male goats, placing each with a sheep mother—a parent of a different species. In adulthood, these male goats refuse to copulate with females of their own species; instead they lust after female sheep—particularly those who have a facial structure much like that of their foster mother. Female goats who grow up with a sheep mother are more flexible in adulthood. Some 50 percent of their mate selections are male sheep

(like their parent figure); but the balance of their amorous overtures are directed toward male goats, individuals of their own species.

6. Judith Rich Harris (1998) presented this unusual hypothesis that one's childhood friends play a larger role in forming one's personality and romantic inclinations than one's parents. This argument has Darwinian logic. For millions of years, young men needed to establish enduring bonds with peers—those who would hunt with them for many years—while young women needed to build ties with other young women who could accompany them for years as they gathered vegetables and help them with their young. Parents would help; but a young adult's relationships with peers would be particularly valuable to his or her long-term survival. So it is logical that teenagers often regard their ties to peers as more important than their ties to parents.

7. In a sample of Americans and Israelis, 47 percent of women and 35 percent of men said that knowing that their suitor was attracted to them "played an important role" in their initial feelings of passion for him or her (Pines 2005). This response most likely emerged during human evolution. When you feel loved by someone you admire, you feel good, and feeling good aids health and longevity.

APPENDIX:
THE MATE CHOICE STUDY

Who We Choose

This book is based on academic studies of romantic attraction, attachment and personality, and on several studies of the anonymous members of Chemistry.com. My Personality Type Study analyzed the responses of 39,913 members to the personality questionnaire found in chapter 2. I discuss most of these results in the book. But I have not included the supporting statistical data in an appendix. I also studied several other attitudes, phenomena and traits by personality type, including income level; level of education; occupation; birth order; types of friends; interests; word usage; political views; religious persuasions; how each type doodles; where each type prefers to live; hand digit length; and beliefs about sex, love and commitment. Although all of these findings are also discussed in the book, I did not include these supporting charts and tables either. All, however, will be available in forthcoming academic articles.

The charts on the following pages show the results of the Mate Choice Study, based on a sample of 28,128 anonymous members of Chemistry.com; specifically, the study identifies the choices made by these individuals when they decided whom to meet for a first date. Of this sample, 17,776 responses were men choosing women and 10,352 responses were women choosing men. Further, 12.2 percent of men in this study were Explorers while 14.5 percent of women were Explorers; 35.4 percent of men were Builders while 37.6 percent of women were Builders; 30.0 percent of men were Directors while 23.4 percent of women were Directors; 22.3 percent of men were Negotiators while 24.5 percent of women were Negotiators. Significantly more men were Directors and more women were Negotiators—a pattern that occurs in every study I made of these four personality types. Each chart tells a different story about human mate choice.

FEMALE EXPLORERS: Who They Choose.

Female Explorers are dramatically drawn to male Explorers (21 percent more than expected by chance). They show no statistically significant likes or dislikes for any other personality type. This may be because Explorers are flexible, optimistic and willing to try just about anything once.

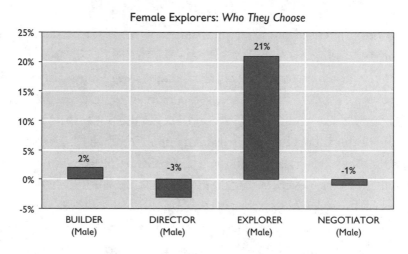

Female Explorers: *Who They Choose*

MALE EXPLORERS: Who They Choose.

Male Explorers are particularly drawn to female Explorers (16 percent more than expected by chance). They show no statistically significant attraction to or dislike for any other personality type. Again, this may be because Explorers are curious and novelty-seeking.

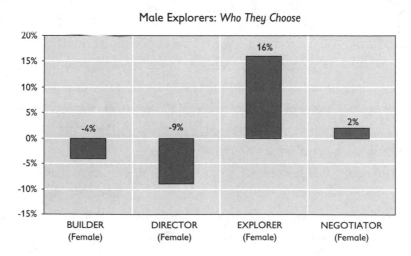

Male Explorers: *Who They Choose*

FEMALE BUILDERS: Who They Choose.

Female Builders are particularly drawn to male Builders (10 percent more than expected by chance). They show no statistically significant attraction to or dislike for Explorers. But they avoid male Directors, perhaps because Builders tend to be reverential and community-oriented while Directors are irreverent and independent. Female Builders also avoid male Negotiators. This may be because Builders are literal, rule-oriented and pragmatic while Negotiators are metaphorical, flexible and idealistic.

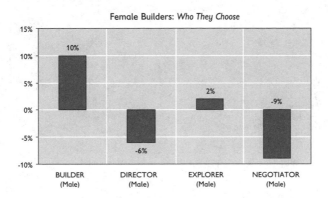

MALE BUILDERS: Who They Choose.

Male Builders are particularly drawn to female Builders (13 percent more than expected by chance). They show no statistically significant attraction to or dislike for Explorers. But they avoid female Directors, perhaps because Builders are social and socially skilled while Directors are autonomous and forthright. They also avoid female Negotiators, perhaps because Builders are factual, respectful of authority and traditional while Negotiators are theoretical, dismissive of authority and unconventional.

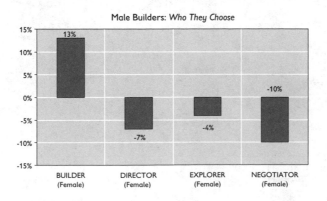

FEMALE DIRECTORS: Who They Choose.

Female Directors have a considerable attraction to male Negotiators (12 percent more than expected by chance). They show no statistically significant preference or dislike for any other type. Perhaps they are equally willing to meet any of these types because they are tough-minded and easily walk away from unsatisfactory introductions.

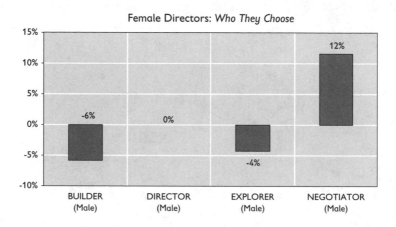

MALE DIRECTORS: Who They Choose.

Male Directors have a significant attraction to female Negotiators (13 percent more than expected by chance). They show no statistically significant attraction to or dislike for female Directors or Explorers. But they avoid female Builders, perhaps because Directors are unconventional and bold while Builders are traditional and cautious.

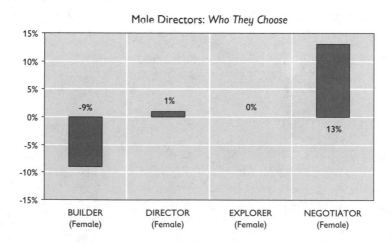

FEMALE NEGOTIATORS: Who They Choose.

Female Negotiators are particularly attracted to male Directors (11 percent more than expected by chance). They show no statistically significant attraction to or dislike for male Negotiators. But they avoid male Builders, perhaps because they are intuitive, flexible and theoretical while Builders are literal, rule-following and factual. Female Negotiators also appear to avoid male Explorers. But there are not enough Negotiator/Explorer pairs in this sample to establish this as a dependable effect. If further research shows this to be a dependable effect, it may be because male Explorers remind female Negotiators of the "bad" boys they knew in high school and college.

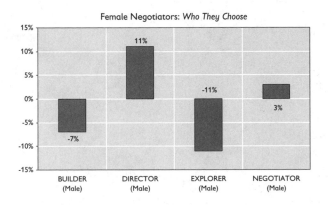

MALE NEGOTIATORS: Who They Choose.

Male Negotiators are particularly attracted to female Directors (15 percent more than expected by chance). Male Negotiators show no statistically significant attraction to or dislike for female Negotiators or Explorers, but they avoid female Builders. Perhaps this is because male Negotiators are theoretical, unconventional and dislike schedules and routines while female Builders are factual, traditional and believe that schedules and routines keep life orderly and relaxing.

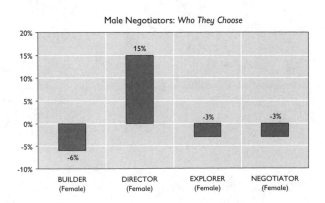

BIBLIOGRAPHY

Aceveda, B., A. Aron, H. Fisher and L. L. Brown (2008). *Neural Correlates of Long-Term Pair-Bonding in a Sample of Intensely In-Love Humans.* Program No. 17912, Washington, D.C.: Society for Neuroscience.

Adorno, T. W., E. Frenkel-Brunswick and D. J. Levinson (1950). *The Authoritarian Personality.* New York: Harper.

Agronick, G. S., and L. E. Duncan (1998). "Personality and Social Change: Individual Differences, Life Path, and Importance Attributed to the Women's Movement." *Journal of Personality and Social Psychology* 74:1545–55.

Aharon, I., N. Etcoff, D. Ariely, C. F. Chabris, O. Nathon (2001). "Beautiful Faces Have Variable Reward Value: fMRI and Behavioral Evidence." *Neuron* 32:537–51.

Ahern, F. M., R. C. Johnson, J. R. Wilson, G. E. McClearn and S. G. Vandenberg (1982). "Family Resemblance in Personality." *Behavior Genetics* 12:261–80.

Ainsworth, M. D. S., M. C. Blehar, E. Waters and S. Wall (1978). *Patterns of Attachment: A Psychological Study of the Strange Situation.* Hillsdale, N.J.: Erlbaum.

Alexander, G. M., and M. Hines (2002). "Sex Differences in Response to Children's

Toys in Nonhuman Primates (*Cercopithecus aethiops sabaeus*)." *Evolution and Human Behavior* 23(6):467–79.

Alford, J. R., C. L. Funk and J. R. Hibbing (2005). "Are Political Orientations Genetically Transmitted?" *American Political Science Review* 99:153–67.

Alicke, M. D., R. H. Smith and M. L. Klotz (1986). "Judgments of Physical Attractiveness: The Roles of Faces and Bodies." *Personality and Social Psychology Bulletin* 12:381–89.

Allik, J., and R. R. McCrae (2004). "Toward a Geography of Personality Traits: Patterns of Profiles across 36 Cultures." *Journal of Cross Cultural Psychology* 35:13–28.

Altman, L. K. (1999). "So, Is This Why Einstein Was So Brilliant?" *New York Times,* June 18.

Aron, A., H. Fisher, D. J. Mashek, G. Strong, H. F. Li and L. L. Brown (2005). "Reward, Motivation and Emotion Systems Associated with Early-Stage Intense Romantic Love: An fMRI Study." *Journal of Neurophysiology* 94:327–37.

Azim, E., D. Mobbs, B. Jo, V. Menon and A. L. Reis (2005). "Sex Differences in Brain Activation Elicited by Humor." *Proceedings of the National Academy of Sciences* 102(45):16496–501.

Bachner-Melman, R., C. Dina, A. H. Zohar, N. Constantini, E. Lerer, S. Hoch, S. Sella, L. Nemanov, I. Gritsenko, P. Lichtenberg, R. Granot and R. P. Ebstein (2005). "$AVPR_{12}$ and $SLC6A_4$ Gene Polymorphisms Are Associated with Creative Dance Performance." *PloS Genetics* 1(3):42.

Backman C. W. (1981). "Attraction in Interpersonal Relationships." In *Sociological Perspective in Social Psychology*, ed. R. Turner and M. Rosenberg. New York: Basic Books, 235–68.

Banissy, M. J., and J. Ward (2007). "Mirror-Touch Synesthesia Is Linked with Empathy." *Nature Neuroscience* 10:815–16.

Baron-Cohen, S. (1995). *Mindblindness: An Essay on Autism and Theory of Mind.* Cambridge, Mass.: MIT Press.

——— (2003). *The Essential Difference: The Truth about the Male and Female Brain.* New York: Basic Books.

Baron-Cohen, S., R. C. Knickmeyer and M. K. Belmonte (2005). "Sex Differences in the Brain: Implications of Explaining Autism." *Science* 310:819–23.

Barr, C. S., T. K. Newman, M. L. Becker, C. C. Parker, M. Champoux, K. P. Lesch, D. Goldman, J. S. Suomi and J. D. Higley (2003). "The Utility of the Non-Human Primate Model for Studying Gene by Environment Interactions in Behavioral Research." *Genes, Brain and Behavior* 2(6):336–40.

Bayne, R. (2005). *Ideas and Evidence: Critical Reflections of MBTI Theory and Practice.* Gainesville, Fla.: Center for the Applications of Psychological Type.

Beach, S. R. H., D. Whitaker, D. J. Jones and A. Tesser (2001). "When Does Performance Feedback Prompt Complementarity in Romantic Relationships?" *Personal Relationships* 8:232–48.

Beall, A. E., and R. J. Sternberg (1995). "The Social Construction of Love." *Journal of Social and Personal Relationships* 12:417–38.

Bell, A. M. (2007). "Future Directions in Behavioural Syndromes Research." *Proceedings of the Royal Society B: Biological Sciences* 274(1611):755–61.

Benbow, C. P., and J. C. Stanley (1983). "Sex Difference in Mathematical Reasoning Ability: More Facts." *Science* 222:1029–31.

Benjamin J., L. Li, C. Patterson, B. D. Greenberg, D. L. Murphy and D. H. Hamer (1996). "Population and Familial Association between the D4 Dopamine Receptor Gene and Measures of Novelty Seeking." *Nature Genetics* 12:81–84.

Ben Zion, I. Z., R. Tessler, L. Cohen, E. Lerer, Y. Raz, R. Bachner-Melman, I. Gritsenko, L. Nemanov, A. H. Zohar, R. H. Belmaker, J. Benjamin and R. P. Ebstein (2006). "Polymorphisms in the Dopamine D4 Receptor Gene (DRD4) Contribute to Individual Differences in Human Sexual Behavior: Desire, Arousal and Sexual Function." *Molecular Psychiatry* 11:782–86.

Berscheid, E., K. K. Dion, E. Walster and G. W. Walster (1971). "Physical Attractiveness and Dating Choice: A Test of the Matching Hypothesis." *Journal of Experimental Social Psychology* 7:173–89.

Berscheid, E., and E. Hatfield-Walster (1978). *Interpersonal Attraction.* 2nd ed. New York: Random House.

Bess, T. L., and R. J. Harvey (2002). "Bimodal Score Distributions and the Myers-Briggs Type Indicator: Fact or Artifact?" *Journal of Personality Assessment* 78(1):176–86.

Biswas, N. M., R. Mazumder, S. K. Bhattacharya and T. K. Das (1985). "Brain 5-hydroxytryptamine and Plasma Testosterone in L-tryptyophan Treated Rats." *Endocrine Research* 11(3–4):131–37.

Blau, P. (1964). *Exchange and Power in Social Life.* New York: John Wiley.

Block, J. (1981). "Some Enduring and Consequential Structures of Personality." In *Further Explorations in Personality,* ed. A. I. Rabin, J. Aronoff, A. M. Barclay and R. A. Zucker. New York: Wiley-Interscience, 27–43.

Bonson, K. R., R. G. Johnson, D. Fiorella, R. A. Rabin and J. C. Winter (1994). "Serotonergic Control of Androgen-Induced Dominance." *Pharmacology Biochemistry and Behavior* 49(2):313–22.

Booth, A., and J. M. Dabbs (1993). "Testosterone and Men's Marriages." *Social Forces* 72:463–77.

Booth, A., G. Shelley, A. Mazur, G. Tharp and R. Kittok (1989). "Testosterone, and Winning and Losing in Human Competition." *Hormones and Behavior* 23:556–71.

Borg, J., B. Andree, H. Soderstrom and L. Farde (2003). "The Serotonin System and Spiritual Experiences." *American Journal of Psychiatry* 160:1965–69.

Botwin, M. T., D. M. Buss and T. K. Shackelford (1997). "Personality and Mate Preferences: Five Factors in Mate Selection and Marital Satisfaction." *Journal of Personality* 65:107–36.

Bouchard, T. J. (1994). "Genes, Environment, and Personality." *Science* 264:1700–1701.

Bouchard, T. J., D. T. Lykken, M. McGue, N. L. Segal and A. Tellegen (1990). "Sources of Human Psychological Differences: The Minnesota Study of Twins Reared Apart." *Science* 250: 223–28.

Bowen, M. (1978). *Family Therapy in Clinical Practice*. New York: Jason Aronson.

Bower, B. (1996). "Gene Tied to Excitable Personality." *Science News* 149:4.

Bowlby, J. (1969). *Attachment and Loss*. Vol. 1, *Attachment*. New York: Basic Books.

Boyle, G. J. (1995). "Myers-Briggs Type Indicator (MBTI): Some Psychometric Limitations." *Australian Psychologist* 30(1):71–74.

Braeutigam, S., S. P. R. Rose, S. J. Swithenby and T. Ambler (2004). "The Distributed Neuronal Systems Supporting Choice-Making in Real-Life Situations: Differences between Men and Women When Choosing Groceries Detected Using Magnetoencephalography." *European Journal of Neuroscience* 20(1):293–302.

Brinberg, E., and P. A. Castel (1982). "A Resource Theory Approach to Interpersonal Interactions: A Test of Foa's Theory." *Journal of Experimental Social Psychology* 43:260–69.

Brisendine, L. (2006). *The Female Brain*. New York: Morgan Roads Books/Random House.

Brosnan, M. J. (2006). "Digit Ratio and Faculty Membership: Implications for the Relationship between Prenatal Testosterone and Academia." *British Journal of Psychology* 97(4):455–66.

Brown, W. M., L. Cronk, K. Grochow, A. Jacobson, C. K. Liu, Z. Popovic and R. Trivers (2005). "Dance Reveals Symmetry Especially in Young Men." *Nature* 438:1148–50.

Brunner, D., and R. Hen (1997). "Insights into the Neurobiology of Impulsive Behavior from Serotonin Receptor Knockout Mice." *Annals of the New York Academy of Sciences* 836(1):81–105.

Burleson, B. R., A. W. Kunkel and J. B. Szolwinski (1997). "Similarity in Cognitive

Complexity and Attraction to Friends and Lovers." *Journal of Contructivist Psychology* 10:221–48.

Buss, D. M. (1984). "Marital Assortment for Personality Dispositions: Assessment with Three Different Data Systems." *Behavior Genetics* 14:111–23.

——— (1985). "Human Mate Selection." *American Scientist* 73:47–51.

——— (1991). "Evolutionary Personality Psychology." *Annual Review of Psychology* 42:459–91.

——— (1996). "Social Adaptation and Five Major Factors of Personality." In *The Five-Factor Model of Personality: Theoretical Perspectives*, ed. J. S. Wiggins. New York: Guilford Press, 180–207.

——— (2003). *The Evolution of Desire: Strategies of Human Mating.* Revised and expanded edition. New York: Basic Books.

Buss, D., and M. Barnes (1986). "Preferences in Human Mate Selection." *Journal of Personality and Social Psychology* 50:559–70.

Buss, D. M., and D. P. Schmitt (1993). "Sexual Strategies Theory: An Evolutionary Perspective of Human Mating." *Psychological Review* 100:204–32.

Byrne, D. (1997). "An Overview (and Underview) of Research and Theory within the Attraction Paradigm." *Journal of Social and Personal Relationships* 14:417–31.

Calvert, J. D. (1988). "Physical Attractiveness: A Review and Reevaluation of Its Role in Social Skill Research." *Behavioral Assessment* 10:29–42.

Campbell, A. (1993). *Men, Women and Aggression.* New York: Basic Books.

——— (2008). "The Morning After the Night Before: Affective Reactions to One-Night Stands among Mated and Unmated Women and Men." *Human Nature* 19(2):157–73.

Canli, T., J. E. Desmond, Z. Zhoa and J. D. E. Gabrieli (2002). "Sex Differences in the Neural Basis of Emotional Memories." *Proceedings of the National Academy of Sciences* 99(16):10789–794.

Cappella, J. N., and J. O. Green (1984). "The Effects of Distance and Individual Differences in Arousability on Nonverbal Involvement: A Test of Discrepancy Arousal Theory." *Journal of Nonverbal Behavior* 8:259–86.

Capraro, R. M., and M. M. Capraro (2002). "Myers-Briggs Type Indicator Score Reliability across Studies: A Meta-Analytic Reliability Generalization Study." *Educational and Psychological Measurement* 62(4):590–602.

Carey, B. (2006). "Living on Impulse." *New York Times,* April 4.

Carter, C. S. (1998). "Neuroendocrine Perspectives on Social Attachment and Love." *Psychoneuroendocrinology* 23:779–818.

Carter, R. (1999). *Mapping the Mind*. Los Angeles: University of California Press.

Carter, S., and C. Snow (2004). "Helping Singles Enter Better Marriages Using Predictive Models of Marital Success." Poster presented at the 16th annual convention of the American Psychological Society. Chicago, Ill., May 27–30.

Cashdan, E. (1995). "Hormones, Sex and Status in Women." *Hormones and Behavior* 29:354–66.

Caspi, A., and E. S. Harbener (1990). "Continuity and Change: Assortative Marriage and the Consistency of Personality in Adulthood." *Journal of Personality and Social Psychology* 58:250–58.

Caspi, A., B. W. Roberts and R. L. Shiner (2005). "Personality Development: Stability and Change." *Annual Review of Psychology* 56:453–84.

Clarke, A. C. (1952). "An Examination of the Operation of Propinquity as a Factor in Mate Selection." *American Sociological Review* 27:17–22.

Cloninger, C. R. (1986). "A Unified Biosocial Theory of Personality and Its Role in the Development of Anxiety States." *Psychiatric Developments* 3:167–226.

——— (1987). "A Systematic Method of Clinical Description and Classification of Personality Variants." *Archives of General Psychiatry* 44(6):573–88.

Cloninger, C. R., T. R. Przybeck and D. M. Svrakic (1991). "The Tridimensional Personality Questionnaire: U.S. Normative Data." *Psychological Reports* 69:1047–57.

Cloninger, C. R., T. R. Przybeck, D. M. Svrakic and R. D. Wetzel (1994). *The Temperament and Character Inventory (TCI): A Guide to Its Development and Use*. St. Louis, Mo.: Washington University, Center for Psychobiology of Personality.

Cohen, E., G. Waugh and K. Place (1989). "At the Movies: An Unobstrusive Study of Arousal Attraction." *Journal of Social Psychology* 129:691–93.

Coleman, M. (1971). "Platelet Serotonin in Disturbed Monkeys and Children." *Clinical Proceedings of the Children's Hospital* 27:187–94.

Costa, P. T., Jr., and R. R. McCrae (1988). "Personality in Adulthood: A Six Year Longitudinal Study of Self Reports and Spouse Ratings on the NEO Personality Inventory." *Journal of Personality and Social Psychology* 54:853–63.

——— (1992). "Appendix A: NEO PI-R Form S Items." In *Revised NEO Personality Inventory (NEO PI-R) and NEO Five-Factor Inventory (NEO-FFI) Professional Manual*. Odessa, Fla.: Psychological Assessment Resources.

——— (1992). "Normal Personality Assessment in Clinical Practice: The NEO Personality Inventory." *Psychological Assessment* 4:5–13.

——— (1992). "Trait Psychology Comes of Age." In *Nebraska Symposium on Motivation: Psychology and Aging*, ed. T. B. Sonderegger. Lincoln: University of Nebraska Press, 169–204.

——— (1995). "Domains and Facets: Hierarchical Personality Assessment Using the Revised NEO Personality Inventory." *Journal of Personality Assessment* 64(1):21–50.

Cramer, R. E., R. G. Cupp and J. A. Kuhn (1993). "Male Attractiveness: Masculinity with a Feminine Touch." *Current Psychology* 12:142–50.

Cravchik, A., and D. Goldman (2000). "Neurochemical Individuality—Genetic Diversity among Human Dopamine and Serotonin Receptors and Transporters." *Archives of General Psychiatry* 57: 1105–14.

Cunningham, M. R. (1986). "Measuring the Physical in Physical Attractiveness: Quasi-Experiments in Sociobiology of Female Facial Beauty." *Journal of Personality and Social Psychology* 50:925–35.

Curran, J. P., and S. Lippold (1975). "The Effects of Physical Attraction and Attitude Similarity on Attraction in Dating Dyads." *Journal of Personality* 43:528–39.

Dabbs, J. M., and M. G. Dabbs (2000). *Heroes, Rogues and Lovers: Testosterone and Behavior*. New York: McGraw-Hill.

Darwin, C. (1910). *The Expression of Emotion in Man and Animals*. New York: Appleton.

Davidge, K. M., L. Atkinson, L. Douglas, V. Lee, S. Shapiro, J. L. Kennedy and J. H. Beitchman (2004). "Association of the Serotonin Transporter and 5HT1K[Beta] Receptor Genes with Extreme Persistent and Pervasive Aggressive Behaviour in Children. *Psychiatric Genetics* 14(3):143–46.

Day, H. (1968). "Curiosity in School Achievement." *Journal of Educational Psychology* 59:33–43.

Decety, J., and P. L. Jackson (2004). "The Functional Architecture of Human Empathy." *Behavioral and Cognitive Neuroscience Reviews* 3(2):71–100.

Depue, R. A., and P. F. Collins (1999). "Neurobiology of the Structure of Personality: Dopamine, Facilitation of Incentive Motivation, and Extraversion." *Behavioral and Brain Sciences* 22(3):491–569.

De Waal, F. (1982). *Chimpanzee Politics: Power and Sex among Apes*. New York: Harper and Row.

De Zegher, F., G. Van Den Berghe, H. Devlieger, E. Eggermont and J. D. Veldhuis (1993). "Dopamine Inhibits Growth Hormone and Prolactin Secretion in the Human Newborn." *Pediatric Research* 34:642–45.

Ding, Y. C., H. C. Chi, D. L. Grady, A. Morishima, J. R. Kidd, K. K. Kidd, P. Flodman, M. A. Spence, S. Schuck, J. M. Swanson, Y. P. Zhang and R. K. Moyzir (2002). "Evidence of Positive Selection Acting at the Human Dopamine Receptor D4 Gene Locus." *Proceedings of the National Academy of Sciences* 99(1):309–14.

Dion, K. K., E. Berscheid and E. Walster (1972). "What Is Beautiful Is Good." *Journal of Personality and Social Psychology* 24:285–90.

Dion, K. K., and K. L. Dion (1975). "Self-Esteem and Romantic Love." *Journal of Personality* 43:39–57.

———— (1985). "Personality, Gender and the Phenomenology of Romantic Love." In *Review of Personality and Social Psychology*, vol. 6, ed. P. Shaver. Beverly Hills, Calif.: Sage.

Dobs, A. S., A. M. Matsumoto, C. Wang and M. S. Kipnes (2004). "Short-Term Pharmacokinetic Comparison of a Novel Testosterone Buccal System and a Testosterone Gel in Testosterone Deficient Men." *Current Medical Research and Opinion* 5:729–38.

Domes, G., M. Heinrichs, A. Michel, C. Berger and S. C. Herpertz (2007). "Oxytocin Improves 'Mind-Reading' in Humans." *Biological Psychiatry* 61(6):731–33.

Drabant, E. M., A. R. Hariri, A. Meyer-Lindenberg, K. E. Munoz, V. S. Mattay, B. S. Kolachana, M. F. Egan and D. R. Weinberger (2006). "Catechol O-methyltransferase Val[158]Met Genotype and Neural Mechanisms Related to Affective Arousal and Regulation." *Archives of General Psychiatry* 63(12): 1396–1406.

Dryer, D. C., and L. M. Horowitz (1997). "When Do Opposites Attract? Interpersonal Complementarity versus Similarity." *Journal of Personality and Social Psychology* 72:592–603.

Du, J., and E. M. Hull (1999). "Effects of Testosterone on Neuronal Nitric Oxide Synthase and Tyrosine Hydroxylase." *Brain Research* 836(1–2):90–98.

Dutton, D. G., and A. P. Aron (1974). "Some Evidence for Heightened Sexual Attraction under Conditions of High Anxiety." *Journal of Personality and Social Psychology* 30:510–17.

Eaves, L. J., H. J. Eysenck and N. G. Martin (1999). *Genes, Culture, and Personality: An Empirical Approach*. New York: Academic Press.

Ebstein, R. P., O. Novick, R. Umansky, B. Priel, Y. Osher, D. Blaine, E. R. Bennett et al. (1996). "Dopamine D4 (D4DR) Exon III Polymorphism Associated with the Human Personality Trait of Novelty Seeking." *Nature Genetics* 12:78–80.

Eisenman, R. (1987). "Creativity, Birth Order, and Risk Taking." *Bulletin of the Psychonomic Society* 25:87–88.

Ellis, B. J. (1992). "The Evolution of Sexual Attraction: Evaluative Mechanisms in Women." In *The Adapted Mind: Evolutionary Psychology and the Generation of Culture*, ed. J. H. Barkow, L. Cosmedes and S. Tooby. New York: Oxford University Press, 267–88.

Erickson, K. I., S. J. Colcombe et al. (2005). "Selective Sparing of Brain Tissue in Postmenopausal Women Receiving Hormone Replacement Therapy." *Neurobiology of Aging* 26(8):1205–13.

Erikson, E. (1959). *Identity and the Life Cycle.* New York: W. W. Norton.

Espejo, E. F. (1997). "Selective Dopamine Depletion within the Medial Prefrontal Cortex Induces Anxiogenic-Like Effects in Rats Placed on the Elevated Plus Maze." *Brain Research* 762(1–2):281–84.

Etcoff, N. (1999). *Survival of the Prettiest: The Science of Beauty.* New York: Anchor Books.

Eysenck, H. J. (1990). "Biological Dimensions of Personality." In *Handbook of Personality: Theory and Research,* ed. L. A. Pervin. New York: Guilford Press, 244–76.

——— (1990). "Genetic and Environmental Contributions to Individual Differences: Three Major Dimensions of Personality." *Journal of Personality* 58:245–61.

Eysenck, H. J., and J. A. Wakefield (1981). "Psychological Factors as Predictors of Marital Satisfaction." *Advances in Behavior Research and Therapy* 3:151–92.

Eysenck, S. B. G., and H. J. Eysenck (1977). "The Place of Impulsiveness in a Dimensional System of Personality Description." *British Journal of Social and Clinical Psychology* 16:57–68.

Farley, F. H., and S. A. Davis (1977). "Arousal, Personality, and Assortative Mating in Marriage." *Journal of Sex and Marital Therapy* 3:122–27.

Farley, F. H., and C. B. Mueller (1978). "Arousal, Personality, and Assortative Mating in Marriage: Generalizability and Cross-Cultural Factors." *Journal of Sex and Marital Therapy* 4:50–53.

Feinberg, D. R., B. C. Jones, M. J. Law Smith, F. R. Moore, L. M. DeBruine, R. E. Cornwell, S. G. Hillier and D. I. Perrett (2006). "Menstrual Cycle, Trait Estrogen Level, and Masculinity Preferences in the Human Voice." *Hormones and Behavior* 49(2):215–39.

Feingold, A. (1990). "Gender Differences in Effects of Physical Attractiveness on Romantic Attraction: A Comparison across Five Research Paradigms." *Journal of Personality and Social Psychology* 59:981–93.

——— (1992). "Gender Differences in Mate Selection Preference: A Test of the Parental Investment Model." *Psychological Bulletin* 112:125–39.

Feldman, R., A. Weller, O. Zagoory-Sharon and A. Levine (2007). "Evidence for a Neuroendocrinological Foundation of Human Affiliation: Plasma Oxytocin Levels across Pregnancy and the Postpartum Period Predict Mother-Infant Bonding." *Psychological Science* 18(11):965–70.

Fenson L., P. S. Dale, J. S. Reznick, E. Bates, D. J. Thal, S. J. Pethick, M. Tomasello,

C. B. Mervis and J. Stiles (1994). "Variability in Early Communicative Development." *Monographs for the Society for Research in Child Development* 59(5):1–185.

Fink, G., B. E. Sumner, J. K. McQueen, H. Wilson and R. Rosie (1998). "Sex Steroid Control of Mood, Mental State and Memory." *Clinical and Experimental Pharmacology and Physiology* 25(10):764–75.

Fink, G., B. E. H. Sumner, R. Rosie, O. Grace and J. P. Quinn (1996). "Estrogen Control of Central Neurotransmission: Effect on Mood, Mental State, and Memory." *Cellular and Molecular Neurobiology* 16(3):325–44.

Fischer, I. V., M. Zuckerman and M. Neeb (1981). "Marital Compatibility in Sensation Seeking as a Factor of Marital Adjustment." *Journal of Sex and Marital Therapy* 7:60–69.

Fischer, I. V., M. Zuckerman and M. Steinberg (1988). "Sensation Seeking Congruence in Couples as a Determinant of Marital Adjustment: A Partial Replication and Extension." *Journal of Clinical Psychology* 44:803–9.

Fisher, H. E. (2006). "Broken Hearts: The Nature and Risks of Romantic Rejection." In *Romance and Sex in Adolescence and Emerging Adulthood: Risks and Opportunities,* ed. A. Booth and C. Crouter. Mahwah, N.J.: Lawrence Erlbaum Associates, 3–29.

Fisher, H. E. (2006). "The Drive to Love: The Neural Mechanism for Mate Choice." In *The Psychology of Love,* 2nd ed., ed. J. R. Sternberg and M. L. Barnes. New Haven: Yale University Press.

Fisher, H. E. (1998). "Lust, Attraction, and Attachment in Mammalian Reproduction." *Human Nature,* 9(1):23–52.

Fisher, H., A. Aron and L. L. Brown (2006). "Romantic Love: A Mammalian Brain System for Mate Choice." In *The Neurobiology of Social Recognition, Attraction and Bonding,* ed. Keith Kendrick. *Philosophical Transactions of the Royal Society: Biological Sciences* 361:2173–86.

Fisher, H., A. Aron and L. L. Brown (2005). "Romantic Love: An fMRI Study of a Neural Mechanism for Mate Choice." *Journal of Comparative Neurology,* 493:58–62.

Fisher, H., A. Aron, G. Strong, D. J. Mashek, H. Li and L. L. Brown (2005). *Motivation and Emotion Systems Associated with Romantic Love Following Rejection: An fMRI Study.* Program No. 660.7. Washington, D.C.: Society for Neuroscience.

Fisher, H. E., and J. A. Thomson, Jr. (2007). "Lust, Romance, Attachment: Do the Side-Effects of Serotonin-Enhancing Antidepressants Jeopardize Romantic Love, Marriage and Fertility?" In *Evolutionary Cognitive Neuroscience,* ed. S. M. Platek, J. P. Keenan and T. K. Shakelford. Cambridge, Mass.: MIT Press, 245–83.

Flaherty, A. W. (2005). "Frontotemporal and Dopaminergic Control of Idea Generation and Creative Drive." *Journal of Comparative Neurology* 493(1):147–53.

Fletcher, G. J., J. A. Simpson and G. Thomas (2000). "Ideals, Perceptions and Evaluation in Early Relationship Development." *Journal of Personality and Social
Psychology* 79:933–40.

Fletcher, G. J., J. A. Simpson, G. Thomas and L. Giles (1999). "Ideals in Intimate
Relationships." *Journal of Personality and Social Psychology* 76:72–89.

Flexner, J. T. (1969). *Washington: The Indispensable Man.* New York: Little, Brown.

Flint, J., R. Corley, J. C. DeFries, D. W. Fulker, J. A. Gray, S. Miller and A. C. Collins
(1995). "A Simple Genetic Basis for a Complex Psychological Trait in Laboratory Mice." *Science* 269(5229):1432–35.

Flory, J. D., S. B. Manuck, K. A. Matthews and M. F. Muldoon (2004). "Serotonergic Function in the Central Nervous System Is Associated with Daily Ratings
of Positive Mood." *Psychiatry Research* 129(1):11–19.

Foa, E. B., and U. G. Foa (1980). "Resource Theory: Interpersonal Behavior as Exchange." In *Social Exchange: Advances in Theory and Research,* ed. K. J. Gergen,
M. S. Greenberg and R. H. Willis. New York: Plenum, 77–94.

Franken, R. E., K. J. Gibson and P. Mohan (1990). "Sensation Seeking and Self-
Disclosure to Close and Casual Friends." *Personality and Individual Differences*
11:829–32.

Freud, S. (1905 / 1965). *Three Essays on the Theory of Sexuality.* New York: Basic Books.

Fullagar, R. (2003). "Kiss Me." *Nature Australia* 27:74–75.

Furnham, A., J. Moutafi and J. Crump (2003). "The Relationship between the
Revised Neo-Personality Inventory and the Myers-Briggs Type Indicator."
Social Behavior and Personality 31(6): 577–84.

Gainetdinov, R. R., W. C. Wetsel, S. R. Jones, E. D. Levin, M. Jaber and M. G.
Caron (1999). "Role of Serotonin in the Paradoxical Calming Effect of Psychostimulants on Hyperactivity." *Science* 283:397–401.

Gangestad, S. W., and R. Thornhill (1997). "The Evolutionary Psychology of
Extra-Pair Sex: The Role of Fluctuating Asymmetry." *Evolution and Human
Behavior* 18(2):69–88.

Gangestad, S. W., R. Thornhill and R. A. Yeo (1994). "Facial Attractiveness, Developmental Stability, and Fluctuating Asymmetry." *Ethology and Sociobiology*
15:73–85.

Gardner, H. (2006). *Five Minds for the Future.* Boston, Mass.: Harvard Business
School Press.

Gazzola, V., L. Aziz-Zadeh and C. Keysers (2006). "Empathy and the Soma-
totropic Auditory Mirror System in Humans." *Current Biology* 16(18):1824–29.

Gerbing, D. W., S. A. Ahadi and J. H. Patton (1987). "Toward a Conceptualization of Impulsivity: Components across the Behavioral and Self-Report Domains." *Multivariate and Behavioral Research* 22:357–79.

Geschwind, N., and A. M. Galaburda (1985). "Cerebral Lateralization. Biological Mechanisms, Associations and Pathology: 1. A Hypothesis and a Program for Research." *Archives of Neurology* 42:428–59.

Giedd, J. N., J. Blumenthal, N. O. Jeffries, F. X. Castellanos, H. Lui, A. Zijdenbos, T. Paus, A. C. Evans and J. L. Rapoport (1999). "Brain Development during Childhood and Adolescence: A Longitudinal MRI Study." *Nature Neuroscience* 2(10):861–63.

Gilger, J. W. (2000). "Contributions and Promise of Human Behavioral Genetics." *Human Biology* 72(1):229–55.

Gillath, O., M. Mikulincer, G. E. Birnbaum and P. R. Shaver (2008). "When Sex Primes Love: Subliminal Sexual Priming Motivates Relationship Goal Pursuit." *Personality and Social Psychology Bulletin* 34:1057–69.

Gillen, B. (1981). "Physical Attractiveness. A Determinant of Two Types of Goodness." *Personality and Social Psychology Bulletin* 7:277–81.

Gilligan, C. (1982). *In a Different Voice.* Cambridge, Mass.: Harvard University Press.

Goldberb, L. R. (1993). "The Structure of Personality Traits: Vertical and Horizontal Aspects." In *Studying Lives through Time: Personality and Development*, ed. D. C. Funder, R. Parke, C. Tomlinson-Keasey and K. Widaman. Washington, D.C.: American Psychological Association, 169–88.

Golimbet, V. E., M. V. Alfimova and N. G. Mityushina (2004). "Polymorphism of the Serotonin 2A Receptor Gene (5HTR2A) and Personality Traits." *Molecular Biology* 38(3):337–44.

Golimbet, V. E., M. V. Alfimova, T. V. Sherbatikh and E. I. Rogaev (2003). "Serotonin Transporter Gene Polymorphism and Personality Traits Measured by MMPI." *Russian Journal of Genetics* 39(4):435–39.

Goodwin, R. (1990). "Sex Differences among Partner Preferences: Are the Sexes Really Very Similar?" *Sex Roles* 23:501–13.

Goreman, B. S., and A. E. Wesman (1974). "The Relationships of Cognitive Styles and Moods." *Journal of Clinical Psychology* 30:18–36.

Gosling, S. D. (2001). "From Mice to Men: What Can We Learn about Personality from Animal Research?" *Psychological Bulletin* 12731:45–86.

Gosling, S. D., and O. P. John (1999). "Personality Dimensions in Non-Human Animals: A Cross-Species Review." *Current Directions in Psychological Science* 8:69–75.

Gosling, S. D., P. J. Rentfrow and W. B. Swann Jr. (2003). "A Very Brief Measure of the Big-Five Personality Domains." *Journal of Research in Personality* 37:504–28.

Gosling, S. D., and S. Vazire (2002). "Are We Barking Up the Right Tree? Evaluating a Comparative Approach to Personality." *Journal of Research in Personality* 36:607–14.

Grimshaw, G. M. (1995). "Relations between Prenatal Testosterone and Cerebral Lateralization in Children." *Neuropsychology* 9(68):74–75.

Grube, J. W., R. R. Kleinhesselink and K. A. Kearney (1982). "Male Self-Acceptance and Attraction toward Women." *Personality and Social Psychology Bulletin* 8:107–12.

Hadjistavropoulos, T., and M. Genest (1994). "The Underestimation of the Role of Physical Attractiveness in Dating Preferences: Ignorance or Taboo?" *Canadian Journal of Behavioral Science* 26:298–318.

Hall, J. A. (1984). *Nonverbal Sex Differences: Communication Accuracy and Expressive Style*. Baltimore: Johns Hopkins University Press.

Halpern, D. F. (1992). *Sex Differences in Cognitive Abilities*. Hillsdale, N.J.: Lawrence Erlbaum Associates.

Ham, B. J., Y. H. Kim, M. J. Choi, J. H. Cha, K. Y. Choi and M. S. Lee (2004). "Serotonergic Genes and Personality Traits in the Korean Population." *Neuroscience Letters* 354(1):2–5.

Hamilton, E. (1942). *Mythology*. Boston, Mass.: Little, Brown.

Hamilton, W. D., and M. Zuk (1982). "Heritable True Fitness and Bright Birds: A Role for Parasites?" *Science* 218:384–87.

Hampson, E. (1990). "Variations in Sex-Related Cognitive Abilities across the Menstrual Cycle." *Brain and Cognition* 14:26–43.

Harmon, A. (2006). "That Wild Streak? Maybe It Runs in the Family." *New York Times,* June 15.

Harris, J. A., J. P. Rushton, E. Hampson and D. N. Jackson (1996). "Salivary Testosterone and Self-Report Aggressive and Pro-Social Personality Characteristics in Men and Women." *Aggressive Behavior* 22:321–31.

Harris, J. R. (1999). *The Nurture Assumption: Why Children Turn Out the Way They Do*. New York: Touchstone.

Haselton, M. G., and D. M. Buss (2000). "Error Management Theory: A New Perspective on Biases in Cross-Sex Mind Reading." *Journal of Personality and Social Psychology* 78(1):81–91.

Hatfield, E., and R. L. Rapson (1993). "Historical and Cross-Cultural Perspectives on Passionate Love and Sexual Desire." *Annual Review of Sex Research* 4:67–98.

Hazan, C., and P. Shaver (1987). "Romantic Love Conceptualized as an Attachment Process." *Journal of Personality and Social Psychology* 52:511–24.

Hedges, L. V., and A. Nowell (1995). "Sex Differences in Mental Test Scores, Variability, and Numbers of High-Scoring Individuals." *Science* 269:41–45.

Heinz A., D. W. Jones, J. G. Gorey, A. Bennet, S. J. Suomi, D. R. Weinberger and J. D. Higley (2003). "Serotonin Transporter Availability Correlates with Alcohol Intake in Non-Human Primates." *Molecular Psychiatry* 8(2):231–34.

Hendrick, S. S., and C. Hendrick (1987). "Multidimensionality of Sexual Attitudes." *Journal of Social and Personal Relationships* 4:281–97.

Heninger, G. R. (1997). "Serotonin, Sex, and Psychiatric Illness." *Proceedings of the National Academy of Sciences* 94:4823–24.

Hensley, W. E. (1994). "Height as a Basis for Interpersonal Attraction." *Adolescence* 29:469–74.

Herlitz, A., and J. E. Yonker (2002). "Sex Differences in Episodic Memory: The Influence of Intelligence." *Journal of Clinical and Experimental Neuropsychology* 24(1):107–14.

Herrmann, D. (1998). *Helen Keller: A Life*. New York: Alfred A. Knopf.

Higley, J. D., P. T. Mehlman, S. B. Higley, B. Fernald, J. Vickers, S. G. Lindell, D. M. Taub, S. J. Suomi and M. Linnoila (1996). "Excessive Mortality in Young Free-Ranging Male Nonhuman Primates with Low Cerebrospinal Fluid 5-hydroxyindoleacetic Acid Concentrations." *Archives of General Psychiatry* 53(6):537–43.

Higley, J. D., and S. J. Suomi (1989). "Temperamental Reactivity in Non-Human Primates." In *Temperament in Childhood*, ed. G. A. Kohnstamm J. E. Bates and M. K. Rothbart. New York: John Wiley, 153–67.

Hinde, R. (1997). *Relationships: A Dialectical Perspective*. Hove, England: Psychology Press.

Holman, T. B. (2001). *Premarital Prediction of Marital Quality or Break Up: Research, Theory, and Practice*. New York: Kluwer.

Houran, J., R. Lange, P. J. Rentfrow and K. H. Bruckner (2004). "Do Online Matchmaking Tests Work? An Assessment of Preliminary Evidence for a Publicized 'Predictive Model of Marital Success.'" *North American Journal of Psychology* 6:507–26.

Houts, R., T. L. Huston and E. Robins (1996). "Compatibility and the Development of Premarital Relationships." *Journal of Marriage and the Family* 58:7–20.

Hoyenga, K. B., and K. T. Hoyenga (1980). *The Question of Sex Differences*. Boston: Little, Brown.

Hoyt, L. L., and J. W. Hudson (1981). "Personal Characteristics Important in Mate Preference among College Students." *Social Behavior and Personality* 9:93–96.

Hsu, M., M. Bhatt, R. Adolphs, D. Tranel, C. F. Camerer (2005). "Neural Systems

Responding to Degrees of Uncertainty in Human Decision-Making." *Science* 310(5754):1680–83.

Hu, X.-Z., R. H. Lipsky, G. Zhu, L. A. Akhtar, J. Taubman, B. D. Greenberg, K. Xu, P. D. Arnold, M. A. Richter, J. L. Kennedy, D. L. Murphy and D. Goldman (2006). "Serotonin Transporter Promoter Gain-of-Function Genotypes Are Linked to Obsessive-Compulsive Disorder." *American Journal of Human Genetics* 78:815–26.

Hughes, S. M., M. A. Harrison and G. G. Gallup Jr. (2007). "Sex Differences in Romantic Kissing among College Students: An Evolutionary Perspective." *Evolutionary Psychology* 5(3):612–31.

Huston, T. L., and R. L. Burgess (1979). "Social Exchange in Developing Relationships: An Overview." In *Social Exchange in Developing Relationships*, ed. T. Huston and R. Burgess. New York: Academic Press, 3–28.

Hyde, J. S., E. Fennema and S. J. Lamon (1990). "Gender Differences in Mathematics Performance: A Meta-Analysis." *Psychological Bulletin* 107:139–55.

Iacoboni, M., I. Molnar-Szakacs, V. Galese and G. Buccino (2005). "Grasping the Intentions of Others with One's Own Mirror Neuron System." *PLoS Biology* 3(3):79–94.

Ickes, W., ed. (1985). *Compatible and Incompatible Relationships*. New York: Springer-Verlag.

Insel, T. R., E. A. Mueller, I. Alterman, M. Linnoila and D. L. Murphy (1985). "Obsessive-Compulsive Disorder and Serotonin: Is There a Connection?" *Biological Psychiatry* 20:1174–88.

Insel, T. R., and L. J. Young (2001). "The Neurobiology of Attachment." *Nature Review of Neuroscience* 2:129–36.

Insel, T. R., J. Zohar, C. Benkelfat and D. L. Murphy (1990). "Serotonin in Obsessions, Compulsions, and the Control of Aggressive Impulses." *Annuals of the New York Academy of Sciences* 600:574–86.

Isaacson, W. (2007). *Einstein: His Life and Universe*. New York: Simon and Schuster.

Jacobs, B. L. (1994). "Serotonin, Motor Activity and Depression-Related Disorders." *American Scientist* 82:456–63.

Jamison, K. R. (1995). *An Unquiet Mind*. New York: Vintage Books.

Jang, K. L., R. R. McCrae, A. Angleitner, R. Riemann and W. J. Livesley (1998). "Heritability of Facet-Level Traits in a Cross-Cultural Twin Sample: Support for a Hierarchical Model of Personality." *Journal of Personality and Social Psychology* 74:1556–65.

Janowsky, J. S., S. K. Oviatt and E. S. Orwoll (1994). "Testosterone Influences Spatial Cognition in Older Men." *Behavioral Neuroscience* 108(2):325–32.

Jasienska, G., S. F. Lipson, P. T. Ellison, I. Thune and A. Ziomkiewicz (2006). "Symmetrical Women Have Higher Potential Fertility." *Evolution and Human Behavior* 27(5):390–400.

Johansson, F., L. Von Knorring and L. Oreland (1983). "Platelet MAO Activity in Patients with Chronic Pain Syndrome: Relationship to Personality Traits, Endorphins in CSF and Plasma Cortisol." *Medical Biology* 61(2):101–4.

Jones, D. (1995). "Sexual Selection, Physical Attractiveness and Facial Neoteny: Cross Cultural Evidence and Implications." *Current Anthropology* 36:723–48.

Jones, E., and K. Hill (1993). "Criteria of Facial Attractiveness in Five Populations." *Human Nature* 4:271–96.

Jung-Beeman, M., E. M. Bowden, J. Haberman, J. L. Frymiare, S. Arambel-Liu, R. Greenblatt, P. J. Reber and J. Kounios (2004). "Neural Activity When People Solve Verbal Problems with Insight." *PLoS Biology* 2:500–510.

Kandel, E. R. (2006). *In Search of Memory: The Emergence of a New Science of Mind.* New York: W. W. Norton.

Karney, B. R., and T. N. Bradbury (1995). "The Longitudinal Course of Marital Quality and Stability: A Review of Theory, Methods and Research." *Psychological Bulletin* 118:3–34.

Keller, H. (2000). *Light in My Darkness,* ed. R. Silverman, West Chester, Pa.: Chrysalis Books.

——— (2003). *The Story of My Life,* ed. R. Shattuck with D. Herrmann. New York: W. W. Norton.

Keller, M. C., and R. K. Young (1996). "Mate Assortment in Dating and Married Couples." *Personality and Individual Differences* 21:217–21.

Keltikangas-Jarvinen, L., M. Elovainio, M. Kivimaki, D. Lichtermann, J. Ekelund and L. Peltonen (2003). "Association between the Type 4 Dopamine Receptor Gene Polymorphism and Novelty Seeking." *Psychosomatic Medicine* 65:471–76.

Kendler, K. S., C. K. Jacobson, C. O. Gardner, N. Gillespie, S. A. Aggen and C. A. Prescott (August 2007). "Genetic Factors Strongly Shape How Peers Are Chosen." *Archives of General Psychiatry* 64(8):958–65.

Kenrick, D. T., G. E. Groth, M. R. Trost and E. K. Sadalla (1993). "Integrating Evolutionary and Social Exchange Perspectives on Relationships: Effects of Gender, Self-Appraisal, and Involvement Level on Mate Selection." *Journal of Personality and Social Psychology* 64:951–69.

Kenrick D. T., E. K. Sadalla, G. E. Groth and M. R. Trost (1990). "Evolution, Traits and the States of Human Courtship: Qualifying the Parental Investment Model." *Journal of Personality* 58(1):97–116.

Kendrick, K. M. (2000). "Oxytocin, Motherhood and Bonding." *Experimental Physiology* 85:111S–124S.

Kerckoff, A. C. (1974). "The Social Context of Interpersonal Attraction." In *Foundations of Interpersonal Attraction*, ed. T. L. Huston. New York: Academic Press, 61–78.

King, J. E., and A. J. Figueredo (1997). "The Five Factor Model plus Dominance in Chimpanzee Personality." *Journal of Research in Personality* 31(2):257–71.

Klohnen, E. C., and G. Mendelsohn (1998). "Partner Selection for Personality Characteristics: A Couple-Centered Approach." *Journal of Personality and Social Psychology* 24:268–78.

Knickmeyer, R., S. Baron-Cohen, P. Raggatt and K. Taylor (2005). "Foetal Testosterone, Social Relationships and Restricted Interests in Children." *Journal of Child Psychology and Psychiatry* 46(2):198–210.

Knickmeyer, R., S. Baron-Cohen, P. Raggatt, K. Taylor and G. Hackett (2006). "Fetal Testosterone and Empathy." *Hormones and Behavior* 49(3):282–92.

Knutson, B., O. M. Wolkowitz, S. W. Cole, T. Chan, E. A. Moore, R. C. Johnson, J. Terpstra, R. A. Turner and V. I. Reus (1998). "Selective Alteration of Personality and Social Behavior by Serotonergic Intervention." *American Journal of Psychiatry* 155(3):373–78.

Kohlberg, L. (1969). "Stage and Sequence: The Cognitive-Developmental Approach to Socialization." In *Handbook of Socialization: Theory and Research*, ed. D. A. Goslin, Chicago: Rand McNally, 347–480.

Kosfeld, M., M. Heinrichs, P. J. Zak, U. Fischbacher and E. Fehr (2005). "Oxytocin Increases Trust in Humans." *Nature* 435:673–76.

Kotulak, R. (1997). "Inside the Brain: Revolutionary Discoveries of How the Mind Works." *Prevention Medicine* 27:246–47.

Kounios, J., J. I. Fleck, D. L. Green, L. Payne, J. L. Stevenson, E. M. Bowden and M. Jung-Beeman (2007). "The Origins of Insight in Resting-State Brain Activity." *Neuropsychologia* 10:1016.

Kposowa, A. J. (2000). "Marital Status and Suicide in the National Longitudinal Mortality Study." *Journal of Epidemiology and Community Health* 54:254–61.

Krueger, R. F., and A. Caspi (1993). "Personality, Arousal, and Pleasure: A Test of Competing Models of Interpersonal Attraction." *Personality and Individual Differences* 14:105–11.

Kurdek, L. A., and J. P. Schmitt (1987). "Partner Homogamy in Married, Heterosexual Cohabiting, Gay, and Lesbian Couples." *Journal of Sex Research* 23:212–32.

Langlois, J. H., L. A. Roggman, R. J. Casey, J. M. Ritter, L. A. Rieser-Danner and

V. Y. Jenkins (1987). "Infant Preferences for Attractive Faces: Rudiments of a Stereotype." *Developmental Psychology* 23:363–69.

Lavrakas, P. J. (1975). "Female Preferences for Male Physique." *Journal of Research in Personality* 9:324–34.

Leckman, J., et al. (1994). "Elevated Cerebrospinal Fluid Levels of Oxytocin in Obsessive-Compulsive Disorder." *Archives of General Psychiatry* 51:782–92.

LeDoux, J. (1996). *The Emotional Brain: The Mysterious Underpinnings of Emotional Life*. New York: Simon and Schuster.

Lepore, F. E. (2001). "Dissecting Genius: Einstein's Brain and the Search for the Neural Basis of Intellect." *Cerebrum: The Dana Forum on Brain Science* 3(1):11–26.

Lerner, R. M., and S. A. Karabenick (1974). "Physical Attractiveness, Body Attitudes, and Self-Concept in Late Adolescents." *Journal of Youth and Adolescence* 3:307–16.

Lesch, K. P., D. Bengel, A. Heils, S. Z. Sabol, B. D. Greenberg, S. Petri, J. Benjamin, C. R. Muller, D. H. Hamer and D. L. Murphy (1996). "Association of Anxiety-Related Traits with a Polymorphism in the Serotonin Transporter Gene Regulatory Region." *Science* 274:1527–31.

Lesnik-Oberstein, M., and L. Cohen (1984). "Cognitive Style, Sensation Seeking and Assortative Mating." *Journal of Personality and Social Psychology* 46:112–17.

Levinger, G. (1986). "Compatibility in Relationships." *Social Science* 71:173–77.

Liechti, M. E., and F. X. Wollenweider (2001). "Which Neuroreceptors Mediate the Subjective Effects of MDMA in Humans? A Summary of Mechanistic Studies." *Human Psychopharmacology* 16(8):589–98.

Lim, M. M., H. P. Nair and L. J. Young (2005). "Species and Sex Differences in Brain Distribution of Corticotropin-Releasing Factor Receptor Subtypes 1 and 2 in Monogamous and Promiscuous Vole Species." *Journal of Comparative Neurology* 487(1):75–92.

Loehlin, J. C. (1992). *Genes and Environment in Personality Development*. Newbury Park, Calif.: Sage.

Loehlin, J. C., R. R. McCrae, P. T. Costa and O. P. John (1998). "Heritabilities of Common and Measure-Specific Components of the Big Five Personality Factors." *Journal of Research in Personality* 32:431–53.

Lorr, M. (1991). "An Empirical Evaluation of the MBTI Typology." *Personality and Individual Differences* 12(11):1141–45.

Luo, S., and E. C. Klohnen (2005). "Assortative Mating and Marital Quality in Newlyweds: A Couple-Centered Approach." *Journal of Personality and Social Psychology* 88(2):304–26.

Lykken, D. T., and A. Tellegen (1993). "Is Human Mating Adventitious or the Result of Deliberate Choice? A Twin Study of Mate Selection." *Journal of Personality and Social Psychology* 65:56–68.

Madsen, D. (1985). "A Biological Property Relating to Power-Seeking in Humans." *American Political Science Review* 79:448–57.

——— (1994). "Serotonin and Social Rank among Human Males." In *The Neurotransmitter Revolution: Serotonin, Social Behavior and the Law*, ed. M. D. Masters and M. T. McGuire. Carbondale: Southern Illinois University Press, 146–58.

Mann, V. A., S. Sasanuma, N. Sakuma and S. Masaki (1990). "Sex Differences in Cognitive Abilities: A Cross-Cultural Perspective." *Neuropsychologia* 28(10):1063–77.

Manning, J. T. (2002). *Digit Ratio: A Pointer to Fertility, Behavior, and Health*. New Brunswick, N.J.: Rutgers University Press.

Manning, J. T., S. Baron-Cohen, S. Wheelwright and G. Sanders (2001). "The 2nd to 4th Digit Ratio and Autism." *Developmental Medicine and Child Neurology* 43:160–64.

Manning, J. T., D. Scutt, G. H. Whitehouse, S. J. Leinster and J. H. Walton (1996). "Asymmetry and Menstrual Cycle in Women." *Ethology and Sociobiology* 17:129–43.

Manuck, S. B., M. E. Bleil, K. L. Petersen, J. D. Flory, J. J. Mann, R. E. Ferrell and M. F. Muldoon (2005). "The Socio-Economic Status of Communities Predicts Variation in Brain Serotonergic Responsivity." *Psychological Medicine* 35(4):519–28.

Manuck, S. B., J. D. Flory, R. E. Ferrell, J. J. Mann and M. F. Muldoon (2000). "A Regulatory Polymorphism of the Monoamine Oxidase-A Gene May Be Associated with Variability in Aggression, Impulsivity, and Central Nervous System Serotonergic Responsivity." *Psychiatry Research* 95(1):9–23.

Manuck, S. B., J. D. Flory, R. E. Ferrell and M. F. Muldoon (2004). "Socio-Economic Status Covaries with Central Nervous System Serotonergic Responsivity as a Function of Allelic Variation in the Serotonin Transporter Gene-Linked Polymorphic Region." *Psychoneuroendocrinology* 29(5):651–68.

Manuck, S. B., J. D. Flory, J. M. McCaffery, K. A. Matthews, J. J. Mann and M. E. Muldoon (1998). "Aggression, Impulsivity, and Central Nervous System Serotonergic Responsivity in a Non-Patient Sample." *Neuropsychopharmacology* 19(4):287–99.

Marioles, N. S., D. P. Strickert and A. L. Hammer (1996). "Attraction, Satisfaction, and Psychological Types of Couples." *Journal of Psychological Type* 26:16–27.

Marks, J. (1999). "How Stimulant Drugs May Calm Hyperactivity." *Science* 283:306.

Martin, N. G., L. J. Eaves, A. C. Heath, R. Jardine, L. M. Feingold and H. J. Eysenck

(1986). "Transmission of Social Attitudes." *Proceedings of the National Academy of Sciences* 83(12):4364–68.

Masters, R. D., and M. T. McGuire, eds. (1994). *The Neurotransmitter Revolution: Serotonin, Social Behavior and the Law.* Carbondale: Southern Illinois University Press.

May, J. L., and P. A. Hamilton (1980). "Effects of Musically Evoked Affect on Women's Interpersonal Attraction and Perceptual Judgment of Physical Attractiveness of Men." *Motivation and Emotion* 4:217–28.

McBride, W. J., J. M. Murphy, L. Lumeng and T. K. Li (1989). "Serotonin and Ethanol Preference." *Recent Developments in Alcoholism* 7:187–209.

McClure, E. B., (2000). "A Meta-Analytic Review of Sex Differences in Facial Expression Processing and Their Development in Infants, Children, and Adolescents." *Psychological Bulletin* 126(3):424–53.

McCrae, R. R. (1987). "Creativity, Divergent Thinking and Openness to Experience." *Journal of Personality and Social Psychology* 52:1258–65.

——— (1996). "Social Consequences of Experiential Openness." *Psychological Bulletin* 120:323–37.

McCrae, R. R., and J. Allik, eds. (2002). *The Five-Factor Model of Personality across Cultures.* New York: Springer-Verlag.

McCrae, R. R., and P. T. Costa (1987). "Validation of the Five-Factor Model across Instruments and Observers." *Journal of Personality and Social Psychology* 52:81–90.

——— (1989). "Reinterpreting the Myers-Briggs Type Indicator from the Perspective of the Five-Factor Model of Personality." *Journal of Personality* 57:17–40.

——— (1997). "Personality Trait Structure as a Human Universal." *American Psychologist* 52:509–16.

——— (2003). *Personality in Adulthood: A Five-Factor Theory Perspective.* New York: Guilford Press.

McCrae, R. R., P. T. Costa, M. Hrebickova, M. E. Kusdil, R. Woodfield, P. R. Saunders, P. B. Smith, F. Ostendorf, A. Angleitner, M. D. Avia, J. Sanz and M. L. Sanchez-Bernardos (2000). "Nature over Nurture: Temperament, Personality and Life Span Development." *Journal of Personality and Social Psychology* 78:173–86.

Meaney, M. (2007). "Environmental Programming of Phenotypic Diversity in Female Reproductive Strategies." *Advances in Genetics* 59:174–90.

Mehrabian, A. (1989). "Marital Choice and Compatibility as a Function of Their Trait Similarity-Dissimilarity." *Psychological Reports* 65:1202.

Melke, J., et al. (2003). "A Polymorphism in the Serotonin Receptor 3A (HTR3A)

Gene and Its Association with Harm Avoidance in Women." *Archives of General Psychiatry* 60(10):1017–23.

Meloy, J. R. and H. E. Fisher (2005). "Some Thoughts on the Neurobiology of Stalking." *Journal of Forensic Sciences,* 50(6):1472–80.

Mendelson, S. D., and B. S. McEwen (1990). "Chronic Testosterone Propionate Treatment Decreases the Concentration of [3H] Quipazine Binding at 5-HT Receptors in the Amygdala of the Castrated Male Rat." *Brain Research* 528(2):339–43.

Meyer-Lindenberg, A., J. W. Buckholtz, B. Kolachana, A. R. Hariri, L. Pezawas, G. Blasi, A. Wabnitz, R. Honea, B. Verchinski, J. H. Callicott, M. Egan, V. Mattay and D. R. Weinberger (2006). "Neural Mechanisms of Genetic Risk for Impulsivity and Violence in Humans." *Proceedings of the National Academy of Sciences* 103(16):6269–74.

Miller, G. (2007). "A Surprising Connection between Memory and Imagination." *Science* 315:312.

Milner, R. (1994). *Charles Darwin: Evolution of a Naturalist.* New York: Facts on File.

Mlot, C. (1998). "Probing the Biology of Emotion." *Science* 280:1005–7.

Moore, M. M. (1985). "Nonverbal Courtship Patterns in Women: Context and Consequences." *Ethology and Sociobiology* 6(4):237–47.

Moreland, R. L., and S. R. Beach (1992). "Exposure Effects in the Classroom: The Development of Affinity among Students." *Journal of Experimental Social Psychology* 28:255–76.

Murakame F., T. Shimomura, K. Kotani, S. Ikawa, E. Nanba and K. Adachi (1999). "Anxiety Traits Associated with a Polymorphism in the Serotonin Transporter Gene Regulatory Region in the Japanese." *Journal of Human Genetics* 44(1):15–17.

Murphy, D. L., C. Wright, M. S. Buchsbaum, A. Nichols, J. L. Costa and R. J. Wyatt (1976). "Platelet and Plasma Amine Oxidase Activity in 680 Normals: Sex and Age Differences and Stability over Time." *Biochemical Medicine* 16:254–65.

Murray, S. L., J. G. Holmes and D. W. Griffin (1996). "The Benefits of Positive Illusions: Idealization and the Construction of Satisfaction in Close Relationships." *Journal of Personality and Social Psychology* 70:79–98.

Murstein, B. (1976). *Who Will Marry Whom.* New York: Springer-Verlag.

Myers, I. B., M. H. McCaulley, N. L. Quenk and A. L. Hammer (1998). *MBTI Manual: A Guide to the Development and Use of the Myers-Briggs Type Indicator.* 3rd ed. Mountain View, Calif.: CPP.

Nagoshi, C. T., R. C. Johnson and F. M. Ahern (1987). "Phenotypic Assortative Mating vs. Social Homogamy among Japanese and Chinese Parents." *Behavior Genetics* 17:474–85.

Neimeyer, G. J. (1984). "Cognitive Complexity and Marital Satisfaction." *Journal of Social and Clinical Psychology* 2:258–63.

Nelson, C. A. (1999). "Neural Plasticity and Human Development." *Current Directions in Psychological Science* 8:42–45.

Netter, P., J. Hennig, B. Meier and S. Rohrmann (1998). "Testosterone as an Indicator of an Altered 5-HT Responsivity in Aggressive Subjects." *European Psychiatry* 13, Suppl. 4:181.

Nettle, D. (2006). "The Evolution of Personality Variation in Humans and Other Animals." *American Psychologist* 61:622–31.

Neyer, F. J., and D. Voigt (2004). "Personality and Social Network Effects on Romantic Relationships: A Dyadic Approach." *European Journal of Personality* 18:279–99.

Nowicki, S., Jr., and E. Yaughn (1999). "Close Relationships and Complementary Interpersonal Styles among Men and Women." *Journal of Social Psychology* 139:473–78.

Nyborg, H. (1994). *Hormones, Sex and Society*. Westport, Conn.: Praeger.

Olson, K. R., C. J. Camp and D. Fuller (1984). "Curiosity and the Need for Cognition." *Psychological Reports* 54:71–74.

Opbroek, A., P. L. Delgado, C. Laukes, C. McGahuey, J. Katsanis, F. A. Moreno and R. Manber (2002). "Emotional Blunting Associated with SSRI-Induced Sexual Dysfunction. Do SSRIs Inhibit Emotional Responses?" *International Journal of Neuropsychopharmacology* 5(2):147–51.

Ozer, D. J., and V. Benet-Martinez (2006). "Personality and the Prediction of Consequential Outcomes." *Annual Review of Psychology* 57:401–21.

Parales-Quenza, C. J., et al. (2006). "Astuteness, Trust and Social Intelligence." *Journal for the Theory of Social Behaviour* 36(1):39–56.

Parks, C. L., P. S. Robinson, E. Sibille, T. Shenk and M. Toth (1998). "Increased Anxiety of Mice Lacking the Serotonin 1A Receptor." *Proceedings of the National Academy of Sciences* 95(18):10734–39.

Peirson, A. R., J. W. Heuchert, L. Thomala, M. Berk, H. Plein and C. R. Cloninger (1999). "Relationship between Serotonin and the Temperament and Character Inventory." *Psychiatry Research* 89(1):29–37.

Peplau, L., and S. Gordon (1985). "Women and Men in Love: Gender Differences in Close Heterosexual Relationships." In *Women, Gender and Social Psychology,* ed. V. O'Leary, R. Unger and B. Wallston. Hillsdale, N.J.: Erlbaum, 275–93.

Persky, H., L. Dreisbach, W. R. Miller, C. P. O'Brien, M. A. Khan, H. I. Lief, N. Charney and D. Straus (1982). "The Relation of Plasma Androgen Levels to Sexual Behaviors and Attitudes of Women." *Psychosomatic Medicine* 44:305–19.

Phelps, E. A. (2004). "Human Emotion and Memory: Interactions of the Amygdala and Hippocampal Complex." *Current Opinion in Neurobiology* 14(2):198–202.

Phillips, K., D. W. Fulker, G. Carey and C. T. Nagoshi (1988). "Direct Marital Assortment for Cognitive and Personality Variables." *Behavior Genetics* 18:347–56.

Pilkington, C. J., A. Tesser and D. Stephens (1991). "Complementarity in Romantic Relationships: A Self-Evaluation Maintenance Perspective." *Journal of Social and Personal Relationships* 8:481–504.

Pittenger, D. J. (2005). "Cautionary Comments Regarding the Myers-Briggs Type Indicator." *Consulting Psychology Journal: Practice and Research* 57(3):210–21.

Plomin, R., and J. Crabbe (2000). "DNA." *Psychological Bulletin* 126:806–28.

Quammen, D. (2006). *The Reluctant Mr. Darwin: An Intimate Portrait of Charles Darwin and the Making of His Theory of Evolution*. New York: W. W. Norton.

Quenk, N. L. (2000). *Essentials of Myers-Briggs Type Indicator Assessment*. New York: John Wiley.

Raleigh, M. J., G. L. Brammer, M. T. McGuire and A. Yuwiler (1985). "Dominant Social Status Facilitates the Behavioral Effects of Serotonergic Agonists." *Brain Research* 348:274–82.

Raleigh, M. J., and M. T. McGuire (1994). "Serotonin, Aggression and Violence in Vervet Monkeys." In *The Neurotransmitter Revolution: Serotonin, Social Behavior and the Law*, ed. R. D. Masters and M. T. McGuire. Carbondale: Southern Illinois University Press, 129–45.

Reif, A., and K. L. Lesch (2003). "Toward a Molecular Architecture of Personality." *Behavioral Brain Research* 139:74–89.

Reik, T. (1964). *The Need to Be Loved*. New York: Bantam.

Reuter, M., S. Roth, K. Holve and J. Hennig (2006). "Identification of First Candidate Genes for Creativity: A Pilot Study." *Brain Research* 1069(1):190–97.

Richards, L. S., J. A. Wakefield and R. Lewak (1990). "Similarity of Personality Variables as Predictor of Marital Satisfaction: Minnesota Multiple Personality Inventory (MMPI) Item Analysis." *Personality and Individual Differences* 11:39–43.

Rikowski, A., and K. Grammer (1999). "Human Body Odour, Symmetry and Attractiveness." *Proceedings of the Royal Society of London B: Biological Sciences* 266:869–74.

Roberts, B. W., and W. F. DelVecchio (2000). "The Rank-Order Consistency of Personality Traits from Childhood to Old Age: A Quantitative Review of Longitudinal Studies." *Psychological Bulletin* 126:3–25.

Robins, R. W. (2005). "The Nature of Personality: Genes, Culture, and National Character." *Science* 310:62–63.

Robins, R. W., J. L. Tracy, K. Trzesniewski, J. Potter and S. D. Gosling (2001). "Personality Correlates of Self-Esteem." *Journal of Research in Personality* 35(4):463–82.

Roloff, M. E. (1981). *Interpersonal Communication: The Social Exchange Approach.* Thousand Oaks, Calif.: Sage.

Rowland, G. L., G. Fouts and T. Heatherton (1989). "Television Viewing and Sensation Seeking: Uses, Preferences and Attitudes." *Personality and Individual Differences* 10:1003–6.

Ruchkin, V. V., R. A. Koposov, B. Klinteberg, L. Oreland and E. L. Grigorenko (2005). "Platelet MAO-B, Personality, and Psychopathology." *Journal of Abnormal Psychology* 114(3):477–82.

Rushton, P. (1988). "Genetic Similarity, Mate Choice, and Fecundity in Humans." *Ethology and Sociobiology* 9:329–34.

Rutter, M. (2007). "Gene-Environment Interdependence." *Developmental Science* 10(1):12–13.

Ruvolo, A. P., and J. Veroff (1997). "For Better or for Worse: Real-Ideal Discrepancies and the Marital Well-Being of Newlyweds." *Journal of Personal and Social Relationships* 14:223–42.

Rytting M., R. Ware and P. Hopkins (1992). "Type and the Ideal Mate: Romantic Attraction or Type Bias?" *Journal of Psychological Types* 24:3–12.

Schalling, D., G. Edman and M. Asberg (1983). "Impulsive Cognitive Style and the Inability to Tolerate Boredom." In *Biological Basis of Sensation Seeking, Impulsivity and Anxiety*, ed. M. Zuckerman. Hillsdale, N.J.: Erlbaum, 123–45.

Schalling, D., et al. (1984). "Impulsivity, Nonconformity and Sensation Seeking as Related to Biological Markers for Vulnerability." *Clinical Neuropharmacology* 7:746–47.

Schierman, M. J., and G. L. Rowland (1985). "Sensation Seeking and Selection of Entertainment." *Personality and Individual Differences* 6:599–603.

Schmitt, D. P. (2002). "Personality, Attachment and Sexuality Related to Dating Relationship Outcomes: Contrasting Three Perspectives on Personal Attribute Interaction." *British Journal of Social Psychology* 41(4):589–610.

Schmitt, D. P., and D. M. Buss (1996). "Strategic Self-Promotion and Competitor Derogation: Sex and Context Effects on the Perceived Effectiveness of Mate Attraction Tactics." *Journal of Personality and Social Psychology* 70:1185–204.

Schrag, A., and M. Trimble (2001). "Poetic Talent Unmasked by Treatment of Parkinson's Disease." *Movement Disorders* 16(6):1175–76.

Segal, M. W. (1974). "Alphabet and Attraction: Unobtrusive Measure of the Effect of Propinquity in a Field Setting." *Journal of Personality and Social Psychology* 30:654–57.

Senior, C. (2003). "Beauty in the Brain of the Beholder." *Neuron* 38:525–28.

Shaikh, T., and K. Suresh (1994). "Attitudinal Similarity and Affiliation Needs as Determinants of Interpersonal Attraction." *Journal of Social Psychology* 134:257–59.

Shepher, J. (1971). "Mate Selection among Second Generation Kibbutz Adolescents and Adults: Incest Avoidance and Negative Imprinting." *Archives of Sexual Behavior* 1:293–307.

Sheppard, J. A., and A. J. Strathman (1989). "Attractiveness and Height: The role of Status in Dating Preferences, Frequency of Dating, and Perception of Attractiveness." *Personality and Social Psychology Bulletin* 15:617–27.

Shibazaki, K., and K. A. Brennan (1998). "When Birds of Different Feathers Flock Together: A Preliminary Comparison of Intra-Ethnic and Inter-Ethnic Dating Relationships." *Journal of Social and Personal Relationships* 15:248–56.

Siebert, C. (2006). "The Animal Self." *New York Times Magazine,* January 22, 48–57.

Simon, N. G., A. Cologer-Clifford, S. F. Lu, S. E. McKenna and S. Hu (1998). "Testosterone and Its Metabolites Modulate 5HT1A and 5HT1B Agonists' Effects on Inter-Male Aggression." *Neuroscience & Biobehavioral Reviews* 23(2):325–36.

Singh, D. (1993). "Adaptive Significance of Waist-to-Hip Ratio and Female Physical Attractiveness." *Journal of Personality ad Social Psychology* 65:293–307.

——— (2002). "Female Mate Value at a Glance: Relationship of Waist-to-Hip Ratio to Health, Fecundity and Attractiveness." *Neuroendocrinology Letters* 23 Suppl. 4:81–91.

Skuse, D. H., R. S. James, D. V. M. Bishop, B. Coppin, P. Dalton, G. Aamodt-Leeper, M. Bacarese-Hamilton, C. Creswell, R. McGruk and P. A. Jacobs (1997). "Evidence from Turner's Syndrome of an Imprinted X-Linked Locus Affecting Cognitive Function." *Nature* 387:705–8.

Smith J. E., V. A. Waldorf and D. L. Trembath (1990). "Single Male Looking for Thin, Very Attractive . . ." *Sex Roles* 23:675–85.

Snyder, M. (1997). "When Belief Creates Reality: The Self-Fulfilling Impact of First Impressions on Social Interaction." In *Experiencing Social Psychology,* 3rd ed., ed. A. M. Pines and C. Maslach. New York: Alfred A. Knopf, 189–92.

Soltz, S., and G. E. Vaillant (1999). "The Big Five Personality Traits and the Life Course: A 45-Year Longitudinal Study." *Journal of Research in Personality* 33:208–32.

Sostek, A. J., A. M. Sostek, D. L. Murphy, E. B. Martin and W. S. Born (1981). "Cord Blood Amine Oxidase Activities Relate to Arousal and Motor Functioning in Human Newborns." *Life Sciences* 28:2561–68.

Spanier, G. B. (1976). "Measuring Dyadic Adjustment: New Scales for Assessing the Quality of Marriage and Similar Dyads." *Journal of Marriage and the Family* 38:15–28.

Sprecher, S. (2001). "A Comparison of Emotional Consequences of and Changes in Equity over Time Using Global and Domain-Specific Measures of Equity." *Journal of Social and Personal Relationships* 18(4):477–501.

Sprecher, S., and P. Regan (2002). "Liking Some Things (in Some People) More than Others: Partner Preferences in Romantic Relationships and Friends." *Journal of Social and Personal Relationships* 19(4):463–81.

Stephen, W. A., E. Berscheid and E. Walster (1971). "Sexual Arousal and Interpersonal Perception." *Journal of Personality and Social Psychology* 20:93–101.

Sunnafrank, M. (1986). "Predicted Outcome Value during Initial Interactions: A Reformulation of Uncertainty Reduction theory." *Human Communication Research* 13(1):3–33.

Sunnafrank, M., and A. Ramirez (2004). "At First Sight: Persistent Relational Effects of Get-Acquainted Conversations." *Journal of Social and Personal Relationships* 21(3):361–79.

Suomi, S. J. (1987). "Genetic and Maternal Contributions to Individual Differences in Rhesus Monkey Biobehavioral Development." In *Perinatal Development: A Psychobiological Perspective,* ed. N. Krasnegor, E. Blass, M. Hofer and W. Smotherman. New York: Academic Press, 397–420.

——— (1991). "Uptight and Laid-Back Monkeys; Individual Differences in Response to Social Challengers." In *Plasticity of Development,* ed. S. E. Brauth, W. S. Hall and R. J. Dooling. Cambridge, Mass.: MIT Press, 27–56.

——— (1999). "Behavioral Inhibition and Impulsive Aggressiveness: Insights from Studies with Rhesus Monkeys." In *Child Psychology: A Handbook of Contemporary Issues,* ed. L. Balter and C. S. Tamis-Lemonda. Philadelphia: Taylor and Francis, 510–25.

——— (2005). "Aggression and Social Behaviour in rhesus monkeys." *Novartis Foundation Symposium* 268:216–22.

Tavernise, S. (2006). "A Baghdad Commander, Armed with Pink Tulle." *New York Times,* July 4.

Taylor, M., and S. G. Vandenberg (1988). "Assortive Mating for IQ and Personality Due to Propinquity and Personal Preference." *Behavior Genetics* 18:339–45.

Taylor, S. E., L. C. Klein, B. P. Lewis, T. L. Gruenewald, R. A. R. Gurung and J. A.

Updegraff (2000). "Biobehavioral Responses to Stress in Females: Tend-and-Befriend, Not Fight-or-Flight." *Psychological Review* 107:441–29.

Terracciano, A., et al. (2005). "National Character Does Not Reflect Mean Personality Trait Levels in 49 Cultures." *Science* 310:96–100.

Thorne, B. (1993). *Gender Play*. New Brunswick, N.J.: Rutgers University Press.

Thornhill, R., and S. W. Gangestad (1993). "Human Facial Beauty." *Human Nature* 4(3):237–69.

Thornhill, R., and A. P. Moller (1997). "Developmental Stability, Disease and Medicine." *Biological Reviews* 72:497–548.

Thornton, B., R. M. Ryckman and J. A. Gold (1981). "Sensation Seeking as a Determinant of Interpersonal Attraction toward Similar and Dissimilar Others." *Journal of Mind and Body* 2:85–91.

Tooby, H. J., and L. Cosmides (1990). "On the Universality of Human Nature and the Uniqueness of the Individual: The Role of Genetics and Adaptation." *Journal of Personality* 58:17–68.

Townsend, J. M., and T. Wasserman (August 1998). "Sexual Attractiveness: Sex Differences in Assessment and Criteria." *Evolution and Human Behavior* 19:171–91.

Turner, C. W., M. H. Ford, D. W. West and A. W. Meible (1986). "Genetic Influences on Testosterone, Hostility, and Type A Behavior in Adult Male Twins." Paper presented at a meeting of the Eastern Psychological Association, Washington, D.C.

Turri, M. G., S. R. Datta, J. DeFries, N. D. Henderson and J. Flint (2001). "QTL Analysis Identifies Multiple Behavioral Dimensions in Ethological Tests of Anxiety in Laboratory Mice." *Current Biology* 11(10):725–34.

Umberson, D., and M. Houghs (1987). "The Impact of Physical Attractiveness on Achievement and Psychological Well-Being." *Social Psychology Quarterly* 50:227–36.

Volavka, J. (1999). "The Neurobiology of Violence: An Update." *Journal of Neuropsychiatry and Clinical Neurosciences* 11:307–14.

Vollm, B. A., A. N. W. Taylor, P. Richardson, R. Corcoran, J. Stirling, S. McKie, J. F. W. Deakin and R. Elliott (2006). "Neuronal Correlates of Theory of Mind and Empathy: A Functional Magnetic Resonance Imaging Study in a Nonverbal Task." *NeuroImage* 29:90–98.

Wacker, J., M. L. Chavanon and G. Stemmler (2006). "Investigating the Dopaminergic Basis of Extraversion in Humans: A Multilevel Approach." *Journal of Personality and Social Psychology* 91(1):171–87.

Waitzkin, J. (2007). *The Art of Learning: A Journey in the Pursuit of Excellence*. New York: Free Press.

Waller, N., and P. Shaver (June 1994). "The Importance of Nongenetic Influences on Romantic Love Styles: A Twin-Family Study." *Psychological Science* 5(5):268–74.

Walster, E., and E. Berscheid (June 1971). "Adrenaline Makes the Heart Grow Fonder." *Psychology Today,* 47–62.

Walster E., G. Walster and E. Berscheid (1978). *Equity: Theory and Research.* Boston: Allyn and Bacon.

Walster, E., G. W. Walster, J. Piliavin and L. Schmidt (1973). "Playing Hard to Get: Understanding an Elusive Phenomenon." *Journal of Personality and Social Psychology* 26:113–21.

Walther, J. B. (1996). "Computer-Mediated Communications: Impersonal, Interpersonal, and Hyper-Personal Interaction." *Communication Research* 23:3–44.

Was, M., A. Combs and J. N. Combs (1999). *Finding Your Strong Suit: How to Read Your Spouse, Boss, Partner, Lover, Opponent and Yourself.* Los Angeles: Renaissance Books.

Watson, D., B. Hubbard and D. Wiese (2000). "General Traits of Personality and Affectivity as Predictors of Satisfaction in Intimate Relationships: Evidence from Self and Partner Ratings." *Journal of Personality* 68:413–49.

Watson, D., E. Klohnen, A. Casillas, N. S. Simms, J. Haig and D. Berry (2004). "Match Makers and Deal Breakers: Analysis of Assortative Mating in Newlywed Couples." *Journal of Personality* 72:1029–68.

Wedekind, C., et al. (1995). "MHC-Dependent Mate Preferences in Humans." *Proceedings of the Royal Society of London* 260:245–49.

Williams, J., and E. Taylor (2006). "The Evolution of Hyperactivity, Impulsivity and Cognitive Diversity." *Journal of the Royal Society Interface* 3(8):399–413.

Williams, S., R. M. Ryckman, J. A. Gold and E. Lenney (1982). "The Effects of Sensation Seeking and Misattribution of Arousal on Attraction toward Similar or Dissimilar Strangers." *Journal of Research in Personality* 16:217–26.

Wilson, A. N. (1988). *Tolstoy.* New York: W. W. Norton.

Wilson, G. (1983). "Finger Length as an Index of Assertiveness in Women." *Personality and Individual Differences* 4:111–12.

Wilson, W. (1989). "Brief Resolution of the Issue of Similarity versus Complementarity in Mate Selection Using Height Preference as a Model." *Psychological Reports* 65:387–93.

Winch, R. F. (1958). *Mate Selection: A Study of Complementary Needs.* New York: Harper.

Winfrey, Oprah (2005). "How I Got There." *Newsweek,* October 24, 48–49.

Witelson, S. F., D. L. Kigar and T. Harvey (1999). "The Exceptional Brain of Albert Einstein." *Lancet* 353:2149–53.

Wolf, M., G. S. van Doorn, O. Leimar and F. J. Weissing (2007). "Life-History Trade-Offs Favour the Evolution of Animal Personalities." *Nature* 447:581–84.

Wright, W. (1990). *Born That Way: Genes, Behavior, Personality*. New York: Routledge.

Yuwiler, A., G. L. Brammer and K. C. Yuwiler (1994). "The Basics of Serotonin Neurochemistry." In *The Neurotransmitter Revolution: Serotonin, Social Behavior and the Law*, ed. M. D. Masters and M. T. McGuire. Carbondale: Southern Illinois University Press, 37–46.

Zajonc, R. B. (1968). "Attitudinal Effects of Mere Exposure." *Journal of Personality and Social Psychology*. Monograph supplement 9(2):1–27.

Zak, P. J., R. Kurzban and W. T. Matzner (2007). "Oxytocin Is Associated with Human Trustworthiness." *Hormones and Behavior* 48(5):522–27.

Zak, P. J., A. A. Stanton and S. Ahmadi (2007). "Oxytocin Increases Generosity in Humans." *PLoS ONE* 2(11):54–71.

Zentall, S. S., and M. J. Meyer (1987). "Self-Regulation of Stimulation of ADD-H Children during Reading and Vigilance Task Performance." *Journal of Abnormal Child Psychology* 15:519–36.

Zentner, M. R. (2005). "Ideal Mate Personality Concepts and Compatibility in Close Relationships: A Longitudinal Analysis." *Journal of Personality and Social Psychology* 89(2):242–56.

Zuckerman, M. (1979). *Sensation Seeking: Beyond the Optimal Level of Arousal*. Hillsdale, N. J.: Erlbaum.

——— (1994). *Behavioral Expression and Biosocial Basis of Sensation Seeking*. New York: Cambridge University Press.

Zuckerman, M., and M. Neeb (1980). "Demographic Influences in Sensation Seeking and Expressions of Sensation Seeking in Religion, Smoking and Driving Habits." *Personality and Individual Differences* 1:197–206.

ACKNOWLEDGMENTS

Thank you, John Sterling, my editor, for your superb guidance, and Claire McKinney, Emi Ikkanda and everyone else at Henry Holt for your enthusiasm for this book. Thank you, Amanda Urban, my literary agent, for your commitment to this project. I give special thanks to Mandy Ginsberg, Thomas Enraght-Moony, Jim Safka, Adam Hanft, Shar Dubey, Francisco Bonilla, Amy Canaday, Mike Presz, Michelle Fretter and everyone else at Match.com and Chemistry.com for your dedication to this complex investigation of mate choice. Thank you, Jonathan Rich, Heide Island, Dan Marchalik and David Zava for your ideas, research assistance and help in data analysis. I also want to thank several people whose books or ideas had a fundamental impact on my thinking, including Keith Kendrick, David Keirsey, Mike Was, Ayala Malach Pines, Elaine Hatfield, Eva Klohnen and Andy Thomson, as well as friends whose support I have deeply appreciated, including Michael Anderson, Sydney Barrows, Arnold

Brown, Lucy Brown, Paul Cohen, Geoffrey Gilmore, Ursula Goode-nough, Marie Lugano, June Nicholson, Brendan Perreault, Michelle Press, Cathy Saypol, Ed Smith, Barb Smuts, Susan Stautberg, Fred Suffet, Lawrence Van Pelt and Jeff Zeig. I thank Fran Mascia-Lees and the rest of my colleagues at Rutgers for their support. I am indebted to David Sloan, Rob Wallace, Betsy Shuller, Brad La Rosa, Kerry Smith and Martin Bashir at ABC-TV for their interest in my data and ideas. Thank you, Ray Carrol, for your wisdom and humor. And special thanks to Fletcher Hodges for his companionship and enduring interest and aid in completing this project. All errors in this book are, of course, mine.

INDEX

Chesterton, G. K., 117
childhood friends, 158, 235n6
chimpanzees, 33
chunking, 108
Churchill, Winston, 96
Clinton, Bill, 106, 117
Clinton, Hillary, 188
Cloninger, Robert, 228n1
cocaine, 56
Coleridge, Samuel Taylor, 221
commitment, 220
competitiveness, 93–94, 169, 177, 200
COMT gene, 228n2
confidence, 77, 92, 206–7
congenital adrenal hypoplasia
    (CAH), 232n3 (chap. 6)
conscientiousness, 32, 75–76, 79
copulatory gaze, 152, 211
courtship
    active listening and humor in, 216
    and being lovable, 206
    bragging and, 207–8
    Builders and, 152, 197–200, 207–8,
        212, 216
    and casual sex, 219–20
    commitment and, 220
    Directors and, 152–53, 200–202,
        208, 212, 216
    Explorers and, 152, 195–97, 208,
        209, 216
    facial expressions and, 211–13
    five-part pickup and, 214–16
    flirting and, 209–13
    handicapping and, 213
    kissing and, 217–19
    and loom and crouch positions, 214
    Negotiators and, 152, 153, 202–5,
        207–9, 216
    and paying for dinner, 217
    and playing hard to get, 208–9
    self-confidence and, 206–7
    See also dating styles

creativity, 7, 25, 49–51, 72, 170
Cronk, Lee, 151
crouch position, 214
curiosity
    of Explorers, 7, 32, 51–52, 57–58,
        127
    of Negotiators, 110–11, 122

Darwin, Charles
    on adaptability, 54
    agreeableness of, 116
    ambiguous feelings about God,
        109–10
    connection to family and friends,
        112
    as empathetic, 113
    imagination of, 104, 107
    on music, 88
    sexual selection theory of,
        103–4
    verbal aptitude of, 118–19
dating styles
    of Builders, 129–33, 141
    of Directors, 133–36, 141
    of Explorers, 125–29, 141
    of Negotiators, 136–40, 141
    See also courtship
Dawkins, Richard, 104
decisiveness, 8, 27, 84, 90–91
Demeter, 74
depression, 122, 219
dialogike, 35
dihydrotestosterone, 91
Dionysus, 50
directness, 84, 89–90, 176, 225n5
Director-Builder match, 175–78, 191,
    199, 202
DIRECTOR/Builders, 136
Director-Director match, 182–84,
    191, 202
Director-Explorer match, 167–70,
    197, 202

ABOUT THE AUTHOR

HELEN FISHER, PHD, one of the world's leading experts on the nature of romantic love and attachment, is the scientific adviser to Chemistry .com, a division of Match.com. She is the author of four previous books, two of which—*The First Sex* and *The Anatomy of Love*—were *New York Times* Notable Books. A research professor of anthropology at Rutgers University, she lives in New York City.